Essentials of
Scientific Research

Essentials of Scientific Research

A Practical Guide

Devon D. Brewer

Evidence Guides
Seattle

Publisher's Cataloging-in-Publication Data

Names: Brewer, Devon D., author.

Title: Essentials of scientific research : a practical guide / Devon D. Brewer.

Description: Seattle : Evidence Guides, 2020.

Identifiers: LCCN 2020912621 (print) | ISBN 978-0-9986154-0-0 (paperback) | ISBN 978-0-9986154-1-7 (hardcover) | ISBN 978-0-9986154-2-4 (ebook : Kindle)

Subjects: LCSH: Science. | Science--Methodology. | Research. | Skepticism. | Reasoning. | Technical writing. | Handbooks and manuals. | BISAC: SCIENCE / Research & Methodology.

Classification: LCC Q175 .B74 2020 (print) | LCC Q175 (ebook) | DDC 507.1--dc23.

Dedicated to my mentors:

A. Kimball Romney

John J. Potterat

Marc L. Miller

and the memory of Linton C. Freeman

Contents

1

Introduction

... I am working for later generations, writing down some ideas that may be of assistance to them. There are certain wholesome counsels, which may be compared to prescriptions of useful drugs; these I am putting into writing; for I have found them helpful in ministering to my own sores, which, if not wholly cured, have at any rate ceased to spread.
—*Lucius Annaeus Seneca, 4 B.C.–A.D. 65*

... it is a constant fact that the most considerable observations and tricks of skill in all sorts of trades and professions are still unwritten.
—*Gottfried Wilhelm Leibniz, 1688–1690*

... this long history of learning how not to fool ourselves—of having utter scientific integrity—is, I'm sorry to say, something that we haven't specifically included in any particular course.
—*Richard Feynman, 1974*

Purpose of this book

Much of science brims with technical, theoretical, mathematical, and statistical sophistication. The general public and students often find scientific research to be hopelessly complex and incomprehensible. Scientists in one field may feel this way about research in other fields, and sometimes even about research in their own fields. However, despite this sophistication, scientific research is also often riddled with flaws in procedure, logic, and language that ultimately undermine its value.

In this book, I cover many indispensable principles and practices for conducting sound scientific research that are rarely, if ever, taught or mentioned in textbooks. I learned most on my own or informally from other experienced researchers. These principles and practices are based on common sense, logic, and

1

fundamentals shared by all sciences. Indeed, the epigraphs that begin chapters in this book show that scientists, philosophers, and others have recognized many of these lessons for centuries, in some cases since antiquity. Yet modern scientists quite often overlook these basic principles and practices. By collecting them in this book, I hope to create a ready reference and reminder for myself and others.

The pursuit of truth often entails many challenges beyond the difficulties of scientific research itself. The biggest challenges are that jobs and institutions dedicated to science may actually (and ironically) deter scientists from seeking the truth. Doing scientific research well is often an entirely separate matter from occupational success as a scientist and prestige in the eyes of other scientists. *This book is about doing scientific research well.* As such, it is not career guide or a handbook on how to survive graduate school or a scientific job.

This book is not a replacement for a methods or other textbook in any field—it is a supplement. The principles and practices I highlight are universal across empirical scientific fields. Although many themes in the philosophy of science run through this book, I focus on practical advice and situations, not abstract philosophical matters. The chapters about writing do not cover grammar, punctuation, or aspects of good writing in general. Rather, in these chapters, I focus on the content of scientific writing.

While I consider each topic in this book to be an essential for scientific research, this book is not a complete inventory of all the practices and principles for doing quality scientific research. I focus on important topics that are relevant in all or nearly all scientific fields, but there are undoubtedly more essentials than I cover.

This book is for anyone who does or wants to do scientific research, regardless of training or experience. Many of the principles and practices I discuss also can be used by non-scientists as a framework for evaluating scientific research on a particular topic.

My background

My training and experience in scientific research have been interdisciplinary. For 34 years, I have worked extensively across the health and social sciences. I have done scientific research in academia, government, and the private sector, as well as independently and as a consultant. About half of my work has been funded by government agencies, foundations, universities, and companies; the other half of my research has been unfunded or self-funded. I have been deeply involved in all of the research activities and roles described in this book. I consider myself a jack of many scientific trades. I have collaborated with several hundred researchers across the world. In my diverse research projects, I have communicated individually, in person or through correspondence, with thousands of other scientists. Apart from my own work, I read widely across many areas of science, and I have a keen interest in the history and philosophy of science.

I have not always followed every principle and practice I highlight in this book. I learned some of the essentials decades ago and others much more recently. If I could go back in time, I would certainly change some of what I did in research and how I did it. As scientists, we are all works in progress. Many of the tips I give in this book could be viewed as common sense. However, by inclination or training, most researchers lack this admirable trait. Rather, we normally learn some lessons through experience. This book gives you the chance to learn from my and others' experience, saving you from costly accidents and mistakes.

Overview of the book

The first section focuses on the steps and procedures common to research in all scientific fields. The subsequent section is about communicating research, which actually is an essential step in doing research. But because scientific communication involves many topics, I devoted an entire section to it. The third section covers the social dimension of working and interacting with scientists and others in research. We scientists (including social scientists) tend to be less socially aware and adept than others, and the chapters here may help us mitigate these deficiencies. The final section on intellectual and ethical matters may be the most important. The intellectual and ethical challenges in doing scientific research are formidable, and I intend to make them easier to tackle with the advice in these chapters. I do not discuss whether studying a particular topic, using a particular approach, or working for a particular purpose, employer, or sponsor is ethical. The answers to these questions likely vary greatly across individuals. Rather, I focus on ethical matters that arise for scientists doing research, regardless of such details.

Although I ordered the chapters into a mostly meaningful sequence, each chapter can stand alone. It is not necessary to read them in order, and in most chapters, I include references to other chapters.

I wrote the chapters in a generic way. I discuss each topic so that it applies to most or all scientists, regardless of field. As a result, this book includes few examples or references. The book also has no tables, graphs, pictures, statistics, or formulas. In my opinion, examples from one field or another would distract readers not familiar with those fields and obscure the points I make. To get the full benefit of this book, as you read, relate the principles and practices I discuss to your own field(s) and specific research problems. Indeed, actively thinking about how the essentials apply to your particular context is the best way to appreciate their value. The lack of references does not mean that I am the first or only person to suggest particular ideas or advice. To the contrary, very few ideas anymore are truly novel. In the Acknowledgments, I list several researchers who have influenced me strongly on the topics in this book, and surely there are many others who have directly or indirectly shaped the views I emphasize here. When I do quote from or refer to another's work specifically, I include the relevant bibliographic information in the Notes section. The Notes also include web addresses for resources that I mention in chapters.

Even though part of this book is about scientific writing, I use a somewhat informal writing style. This book is long on advice, which I give directly to you, the reader. I also assume that you are a scientist or potential scientist, so sometimes I refer to "we," meaning you, me, and other scientists. Most scientific writing, however, is short on advice and has a more formal style than I use in this book.

Please join me in pursuing the essentials of scientific research!

Doing Research

2

No Substitute for Field Experience and Observation

An ounce of practice is worth a pound of theory.
—Robert Gascoyne-Cecil, 1859

Experience and observation in the field are the foundation of successful research projects, even those not conducted in the field. By "field," I mean the natural setting to which the research ultimately relates. Field experience and observation ideally come before beginning a research project. No amount of intelligence, theoretical knowledge, or methodological skill can outweigh experience and observation in the field. Many research projects have failed or produced meaningless results because the researchers did not have corresponding field experience that could have helped them avoid disaster or irrelevance.

When in the field, attempt to experience your subject holistically, with open eyes, ears, and mind. It is not important to have specific objectives for your observation. Rather, the goal is to increase your familiarity and gain at least an initial appreciation for the subject "on the ground." Your experience in the field may inspire research questions. If you already have identified research questions, your field experience and observation will allow you to assess whether your research questions are relevant and meaningful, and whether other questions should be added or studied instead. Field experience and observation can also show you what research designs and methods are feasible and appropriate, and give you background and insights for interpreting results.

Sometimes, it is not possible to get first-hand experience and observation before starting a research project. In this situation, drawing on others' experiences and observations may need to suffice. Study their accounts of the field as recorded in writing, video, or audio. Seek the reports from multiple observers, and interview them yourself, if possible. Veteran field research support staff and technicians can be especially useful advisors, as can individuals who live or work in the field but have no connection to research.

For some subjects, the "field" can be experienced or observed only indirectly with technology, and sometimes very crudely, or in laboratory or other artificial contexts. In such cases, field experience and observation may by necessity occur only in the context of research. The key is to have that experience and observation before developing a new research project. This experience will give you a sense of the context and may provide a contrast to your book knowledge.

Field experience and observation are essential for all researchers, novice and veteran alike. It is especially critical for researchers studying new substantive topics, researchers whose main interest is using a particular method (regardless of topic), and research managers who oversee but do not do the actual research (see Chapter 38). Even if a project is a collaboration between field scientists and researchers working in a laboratory or office, the latter group still may profit from direct exposure to the field. Their field experience may improve communication with their field colleagues and increase their understanding of the topic under study.

Knowledge gained from field experience and observation sometimes is perishable. Conditions in the field can change. Therefore, even researchers who study the same topic over a long period benefit from regular exposure to the field.

Researchers' connection to a subject of study often consists of just a data set that someone else collected (see Chapter 10). In this situation, scientists benefit from direct exposure to the same or similar specimens, events, artifacts, or other phenomena represented in the data. Seek experienced colleagues to orient you to these ultimate source materials and ask them questions to get a better understanding not only of the data but the underlying phenomena as well. In other cases, observing how the data are collected can reveal deficiencies and limitations that are crucial to know for meaningful analysis of the data. Even the best documentation of a data set may lack potentially relevant details, so any opportunity to learn how the data come to be is worthwhile. To understand what a data set *means* and how it *works*, see how it is *made*.

While field experience and observation are essential before a project begins, they are also useful on an ongoing basis, at any point during or after a project. The field is an infinite and reliable guide for scientific research.

3

Designing and Planning Research

... that which is done can never be undone.
—*Plato, 360 B.C.*

... before undertaking any enterprise, careful preparation must be made.
—*Marcus Tullius Cicero, 44 B.C.*

Hofstadter's Law: It always takes longer than you expect, even when you take into account Hofstadter's Law.
—*Douglas Hofstadter, 1979*

The research process starts with formulating specific research questions (see Chapter 48). These questions can arise from any source, although it is critical that they be meaningful and feasible for study (see Chapter 2).

State research questions explicitly and in open-ended fashion. Such questions typically begin with "what," "when," "where," "how," and "why." Avoid questions that presuppose a "yes" or "no" answer. Such questions (and answers) for most topics in most fields are trivial in scientific terms, reflecting limited binary concepts that reveal little about nature, and they curtail deeper and broader investigation. Yes/no questions can almost always be reframed more meaningfully as open-ended questions. For example, instead of asking "is X present?", ask "how much X is present?"

Research design determines what can be known from your research. Studies are made or broken at the design stage. Even perfect data collection, analysis, and reporting cannot redeem a poorly designed study. Align design components with corresponding research questions. For instance, if a research question focuses on cause and effect, it implies an experimental design (see Chapter 52). Ensure the research setting and sampling approach allow you to generalize in the way you want (see Chapter 21). Strive for comparability with prior research in terms of methods, and particularly measures (see Chapter 14). Your planned research, of course, may involve new or different approaches, but including comparable methods helps build knowledge.

Search the literature extensively before finalizing your design (see Chapter 5). You may discover a line of research that makes a research question irrelevant, or find useful approaches that you can adopt. Although searching the literature is a standard practice for students preparing research proposals, it is not necessarily one for trained scientists, in my experience. Grant and contract proposals often involve abbreviated literature reviews, but they are rarely comprehensive. Some scientists may believe they know the literature and therefore do not need to search it. In most scientific fields, the volume of research and multiplicity of outlets for it are too much for researchers to stay abreast of through informal means. Only through systematic and extensive searches can researchers know the literature.

When developing a research project, attack your own ideas and plans continually. At the scientific level, adopt the views of critics. How would they criticize the logic and methods? What can you do to blunt or eliminate those criticisms? Extend yourself further, and assess each assumption and piece of the project's logic and methods. Play devil's advocate and try to imagine all possible weaknesses. For each weakness or problem you identify, make changes to your design, if possible. Thoroughly examine rival schools of thought, approaching them sympathetically. If your project involves evaluating an hypothesis, seek ways to destroy it with the evidence from the planned research. If possible, design your evaluation so that the results expected by one hypothesis would be different than those expected by competing hypotheses.

At the practical level, try to anticipate obstacles and glitches. Play worst case scenarios in your mind. What would happen if a piece of equipment failed? A key relationship with another researcher or collaborating organization soured? Funding decreased? Project staff left? Access to materials, equipment, participants, field sites, or data disappeared? Assess software and hardware for all types for vulnerabilities and devise backup solutions. Think like a planner of a space exploration mission— prepare for things to go wrong and have alternatives for when they do. Moreover, count on all research activities taking longer —usually much longer—than you anticipate. This is especially true for research that involves elements new to you, including topics, methods, equipment, materials, sites, partners, staff, and sponsors.

Every research plan has potential problems. If you cannot identify many, seek feedback on your plans from others with research experience in a similar area (see Chapter 31). The goal of the negative thinking process is to kill bad ideas and strengthen viable ones. I estimate that 90% or more of the research ideas that I commit to writing I never pursue, usually because they are problematic in scientific or practical terms. The entrepreneurial maxim of "fail early, fail often" also holds for scientific research: by weeding out weak research project ideas in the design phase, researchers can devote more time to better prospects.

Even if problems are not caught or prevented in the design stage, negative thinking is still a useful tool to apply throughout the research process. Helpful questions to ask yourself may include "what can go wrong here?" and "what gaps in logic are there with this assumption, line of reasoning, method, or analytic technique?" We naturally have our blind spots, but repeated self-interrogation can sometimes reveal them, allowing chances to fix problems before they get worse.

4

Pre-Registering Research

Light itself is a great corrective. A thousand wrongs and abuses that are grown in darkness disappear, like owls and bats, before the light of day.
—James Garfield, 1831–1881

Unreliable research is a plague on science. Although pre-registering a study protocol is not a panacea, all scientists can use the practice to improve the quality of their research and turn back this plague. Pre-registration involves specifying all study procedures and hypotheses (if relevant) in a document and archiving it in a public registry or publishing it before beginning a research project. By pre-registering, researchers demonstrate their reported research conforms to plans made in advance and the results they report do not stem from procedural changes, data dredging, or post-hoc hypotheses. Protocol registries also provide a record of attempted studies, which enables researchers to assess publication and reporting biases — that is, how well reported studies (and results in particular studies) represent all research that was done.

Pre-registration enforces discipline and transparency, both essential elements of scientific research. When we scientists develop our pre-registered protocols, we engage in proactive thinking and hone our analytic skills, running all analyses mentally and envisioning potential outcome scenarios. Key data management decisions can be documented in the protocol, before data collection. The protocol development process also may stimulate us to consider alternative explanations and hypotheses, which may lead us to revise our research design and plan. Indeed, most of the intellectual work in research occurs during the design, planning, and protocol development phase (see Chapter 3). Furthermore, the pre-registered protocol serves as a road map for executing the study. Pre-registering even gives you a head start on writing up your research, as sections of the protocol often can be used in reports that describe project results (see Chapters 18 and 19).

In the past, some researchers conducted their research as if they had pre-registered their protocols, adhering to plans made before they began their studies. Unfortunately, it is impossible for others to know which studies they performed in this way and to what extent.

Professors require their students to use a similar process to pre-registration. Before students carry out their thesis or dissertation research, they usually must write a research proposal with clear specification of research questions, hypotheses (if relevant), and methods. Yet professors and other trained scientists typically do not do the same for their own research, unless they seek funding (and even in that case, the proposals lack the crucial details of a protocol and are not archived publicly).

Although pre-registration is a relatively recent innovation, several registries exist (see Notes). The process of pre-registration is easy and fast. When pre-registering a protocol, include as much specific information as possible on planned procedures and activities, perhaps as supplementary documents or other materials. It is also good to cover the background and rationale for the research and the hypotheses to be evaluated, if any.

It is often impossible to anticipate all events and dilemmas that might arise during a research project. Therefore, update the protocol as needed throughout the project in response to unexpected circumstances, problems, and opportunities. You can also revise the protocol to take advantage of more appropriate methods, such as statistical techniques, that you discover after registering the original protocol. Avoid altering the protocol based on any inspection or analysis of the project's data. The key is to document decisions and plans for action before executing them, and thereby preserve the research as hypothesis evaluation or description, as opposed to exploration (see Chapter 51). If you do update or revise your protocol, make sure to keep all versions archived at the registry so that others can see what changes you made and when you made them.

Some scientists and sponsors are reluctant to pre-register research because they fear others will use their ideas, as in publishing first or gaining a commercial, military, or other advantage. Some registries allow researchers to keep their pre-registration private for up to a few years. Researchers (and their sponsors) who worry about losing a competitive advantage even with such embargoes likely do not intend to publish their work at all.

Pre-registration is vital for almost all kinds of research in any field, even if that field has no history of pre-registration. It does not matter whether the research is primary or secondary, experimental or observational, or basic or applied—pre-registration increases its scientific value dramatically. Even if you have already collected the data for your study or casually inspected them, pre-registering your analysis plans still introduces some discipline to your work and reduces threats of bias present in analysis projects that are not pre-registered.

5

Searching the Literature

Nothing is so difficult but that it may be found out by seeking.
—Publius Terentius Afer, 163 B.C.

Our own generation enjoys the legacy bequeathed to it by that which preceded it. We frequently know more, not because we have moved ahead by our own natural ability, but because we are supported by the [mental] strength of others, and possess riches that we have inherited from our forefathers. Bernard of Chartres used to compare us to [puny] dwarfs perched on the shoulders of giants. He pointed out that we see more and farther than our predecessors, not because we have keener vision or greater height, but because we are lifted up and borne aloft on their gigantic stature.
—John of Salisbury, 1159

Facts do not cease to exist because they are ignored.
—Aldous Huxley, 1927

Science ultimately is a collective enterprise. We learn from our predecessors and peers, and we seek to inform others of our discoveries. The main way we learn from other researchers is to search the literature.

By searching the literature, we can get or shape ideas, and find answers to our questions. From the results of searches, we can put our own research in context and give credit to those whose work we have relied on, extended, or challenged. Literature searches allow us to avoid reinventing the wheel and repeating others' failures.

Searching the literature also teaches us important history lessons. Most avenues of research eventually become dead ends, or at least paths no longer taken in the pursuit of truth. Few ideas or empirical results are original. If we take these lessons to heart, we gain humility about our own and others' research today.

The best time to search the literature is before we carry out, or even plan, a research project. Each new publication is like a sentence in the worldwide conversation that we scientists have with each other over time. If we want others to understand the research we report, we must know what the conversation is about, what's already been said, and what might be a useful addition. Of course, we can search the literature at any point when we have a need, but it is essential to search the literature formally before we start new research. Even though we learned this sequence as students, many professional researchers cease to follow it when they no longer must satisfy the requirements of their thesis or dissertation committees.

During most of the 20th century, the best ways to search the literature were to browse the shelves in a library (with related material shelved together), inspect journals' tables of contents, find sources from references cited in articles and books (backward citation search), and use print copies of citation indexes to identify sources citing particular works (forward citation search). Although valuable, these tools were very time-consuming to use.

The landscape changed in the late 1990s with the advent of online databases of journal articles, such as PubMed. These databases focus on subsets of journals in broad academic fields. They store information on bibliographic details, abstracts, and subject classifications, and can be searched easily.

Then, in 2004, Google released Google Scholar, a search engine that indexes the full text of the scholarly literature—journals, books, reports, and other documents. I believe that Google Scholar has become the single best tool for searching the literature. It has many advantages: it is free, returns documents published in any language, enables easy forward citation searches, identifies articles related to documents returned in a search, exports citations to bibliographic software, and often provides links to free, full text versions of the documents identified. To my knowledge, no other current scientific literature database or search engine offers full-text indexing. That means if a search term is not in the abstract, subject classification, or bibliographic fields, but is in the text of the document itself, Google Scholar can find the document, but other databases and search engines will not.

In the past, Google Scholar sometimes had more gaps for relatively old material, but those gaps have been and are steadily being filled as older books and articles are digitized. In fact, other databases and search engines are almost useless for documents published more than 60 years ago, because they do not index abstracts of the small amounts of old material they include. In my experience, no other search engine or database covers old material nearly as well as Google Scholar.

Compared to other databases and search engines, however, Google Scholar produces more irrelevant records in a search, although these tend to be concentrated toward the middle and end of the records returned in a search. A few fields, such as chemistry, are not represented well due to uncooperative publishers. I have found excellent coverage in most fields. Indeed, in recent studies, Google Scholar captured the vast majority of the scientific literature, while other services performed much worse. Even just for journal content, other services are categorically incomplete, routinely excluding letters to the editor, commentaries, and editorials that often contain vital scientific information. Google Scholar includes these kinds of publications.

For almost every reason scientists search the literature, it is essential that the search be as complete as possible. In our own studies, we wouldn't throw away half or more of our data blindly. Similarly, it would be foolish to neglect half or more of the relevant research when searching the literature. With Google Scholar, and a little extra effort, we can obtain a fairly complete picture of the literature. As the situation demands, we can also supplement Google Scholar with other databases and search engines, especially those that focus on particular kinds of documents that don't always appear on the World Wide Web, such as dissertations and theses. And forward and backward citation searches remain critical complementary tools for identifying pertinent work not otherwise findable.

Complete searches include all kinds of scientific reports and communications, regardless of publication status or language. Scientific works published in digital form can be translated, at least roughly, between many languages with online translators. Multilingual colleagues also might be able to help with more accurate translations or translating works not in digital form.

Despite the remarkable advances in tools for searching the literature, I have noticed a remarkable decline in the quality and completeness of literature searches in research reports published in recent decades, including meta-analyses and systematic reviews. Just as it has become very easy to search the literature well, many researchers do not attempt to do so. I do not know the reasons for the decline, but it represents a corrosion of scientific practice and threat to the efficient cumulation of knowledge. As George Santayana observed in 1905, "those who cannot remember the past are condemned to repeat it."

6

Collecting Data

Do not seek for information of which you cannot make use.
—Anna C. Brackett, 1892

... the observer is never entirely replaced by instruments; for if he were, he could obviously obtain no knowledge whatsoever ... the observer must, in using the instrument for his investigation, take readings on it ... the most careful record, when not inspected, tells us nothing.
—Erwin Schrödinger, 1944

The careful recording of all details in experimental work is an elementary but important rule. It happens surprisingly often that one needs to refer back to some detail whose significance one did not realise when the experiment was carried out ... Apart from providing an invaluable record of what is done and what observed, notetaking is a useful technique for prompting careful observation.
—W. I. B. Beveridge, 1950

In most fields, there tends to be a status hierarchy of tasks, roughly following the amount of training required to perform them. Collecting data usually is the lowest status, with increasingly higher status for managing data, analyzing data, and interpreting/designing research. This hierarchy is almost opposite of the hierarchy of importance. Data management, analysis, and interpretation all can be redone or revised easily; data collection cannot. The fate of a study rests on the data. Collecting data is therefore the most important part of any study, apart from designing and planning the study (see Chapter 3). For most projects, data collection also tends to be the activity that lasts longest. Thus, inattention to data collection not only threatens the scientific value of a study, but risks wasted effort, time, and money.

Doing it yourself

As scientists advance in their careers, they typically do fewer low status research tasks than they did before. The most prominent scientists with large research groups often do interpretation and design of research only, if even that (see Chapter 38). Scientists who don't collect, manage, and analyze their own data are handicapped in interpreting them and in designing and planning research. This handicap can be reduced, but not eliminated, in the right circumstances. Their colleagues, staff, and students who do these tasks would need to do them very well, and document and communicate their work extensively. The lead scientist must then review their work in fine detail, listen to the research team carefully, and ask them questions exhaustively. Such an approach is inherently inefficient, hence few lead scientists implement it.

If you involved with collecting the data, you can detect problems with or artifacts in the data that less experienced and less skilled researchers miss. Participating in data collection also gives you the chance to observe unexpected phenomena that may give you insights to interpreting your results or suggest new topics or hypotheses to examine.

Controlling your appetite

Researchers often try to collect all the data they can, even on matters outside the explicit focus of the project. They do so because they might not get the chance to collect the data again or they want to explore other topics and questions at the same time. Collecting too much data imperils the overall project. It can reduce data quality, as resources of all kinds are stretched too thin to ensure reliable and valid data, especially when humans are the instruments or data sources. Planning for the additional data and collecting them can prolong the data collection period, increase the time required for data management, and subtract time for analysis and reporting. Scientists may be especially prone to collecting too much data in grant-funded projects. Researchers who plan to collect a lot of data can propose long and large projects with many staff and big budgets. Much, if not most, of the data in such projects can go unanalyzed. To avoid problems and waste, gather no more data than necessary to address your project's central goals.

Testing and monitoring

Pilot test your data collection procedures in the exact circumstances of your study. If equipment malfunctions or your approach fails for some reason, try to fix the problem and pilot test again until you resolve it. Such testing is especially critical when you use methods or collect data in settings that are new to you.

On a regular basis, check automated data collection instruments to ensure they are functioning normally and monitor any other human data collectors to confirm their compliance with procedures. Inspect incoming data continually for errors and gaps (see Chapter 7). Researchers who neglect monitoring often find they have no or useless data once data collection ends.

Documenting data collection

Collecting data involves more than just data. To interpret data, we scientists need to know details about the context in which they were collected, problems that occurred during data collection, and any changes to procedures, whether observational or experimental. Laboratory scientists and many field researchers have long kept notebooks for recording such information with dated entries. Documenting data collection in notebooks or their equivalents is essential for all researchers, regardless of setting or discipline. These notebooks are vital scientific information and should be backed up (see Chapter 12) and archived publicly (see Chapter 9) just like data. As much as possible, also document data collection with other methods, such as photography and audio/video recording. Capture the data collection process in action, along with the equipment, materials, and settings involved. These recordings help demonstrate that your data are genuine and give good material to illustrate study procedures in reports and presentations.

Staying in sequence

Study activities form a natural one-way sequence: designing and planning → collecting data → analyzing data → interpreting and reporting results. Reverting to a prior activity in the sequence undermines scientific integrity. When collecting data, scientists sometimes get impatient and want to analyze the data, informally or formally, to see how the results are developing. Often they hope to reach an arbitrary and irrelevant statistical threshold (see Chapter 8) for an hypothesized outcome. Premature looks are harmful because they can influence whether researchers continue the study or how they conduct the rest of it. In effect, deviating from the research sequence is tampering with the evidence. Sound research practice is simple: do one activity at a time, and complete it before advancing to the next. Analyze the data *after* you have finished collecting them.

7

Managing Data

The card-player begins by arranging his hand for maximum sense.
Scientists do the same with the facts they gather.
—Isaac Asimov, 1988

Managing data is like what a librarian does in creating and maintaining a collection: preparing material for use, labeling and describing it, and then organizing it for easy access and analysis. Data management usually begins before data collection. Without established procedures for storing and examining data, it can be difficult to monitor data collection and detect errors or other problems (see Chapter 6).

In most projects, researchers must modify the raw computerized data in various ways before they can be analyzed meaningfully (see Chapter 8). We often must account for missing data and other errors (such as out of range values), transform the data mathematically, construct composite variables, classify observations, and/or perform other operations. A key step in this process is to inspect the raw data directly and visually if at all feasible. An analytic summary of the raw data can help detect many errors, but the strategy is not foolproof. By definition, you cannot anticipate unexpected problems with the data, so look at the raw data yourself carefully.

Save, set aside, and back up (see Chapter 12) the raw data prior to any changes. Do the same after each major wave of changes you make to the data, so that you can roll back to a prior state of the data quickly and easily. Record all changes to the data that you make along with the dates and reasons for them in a separate document (and also within the data file itself, if possible). This document may serve as the basis of a data dictionary, codebook, or guide for the data set. When documenting the data, give the full details on their structure and

26

content such that a person not involved with the project can understand precisely what the data represent and any potential issues they pose for analysis. If your project involves many data files, use file names that are descriptive and specific so that they have some meaning if the documentation is not readily available.

Basic tools for data management include spreadsheets, text editors, statistical software packages, programming languages, and relational databases. I recommend learning how to use at least one instance of each type. Although these tools' functions overlap, each tool allows you to do some common data management tasks that the others cannot (or cannot nearly as easily or fast). Even if other researchers in your field tend not to use a particular tool, try learning it anyway, because you will likely discover a function that saves you much time and effort in data management.

In my experience, scientists are least familiar with relational databases, such as MySQL and Microsoft Access. They are especially useful for managing complex data sets. Relational databases are also well suited for collecting data (if it involves automated online data extraction, or humans recording data manually) and monitoring of data collection. Just as with programming (see Chapter 11), relational databases open new windows on the world. Compared to other data management tools, they involve different ways of thinking about data structures and different approaches to solving data management and analysis problems. If your research involves relational databases at any stage, become proficient in developing and modifying them. This will reduce your dependence on database developers and analysts who may not be available when you need them or may charge more than you can afford. If you work with reliable data management staff or vendors, your familiarity with relational databases will help improve your communication with them and allow you to plan your research operations more efficiently.

8

Analyzing Data

... when you can measure what you are speaking about, and express it in numbers, you know something about it; but when you cannot express it in numbers, your knowledge is of a meagre and unsatisfactory kind.
—William Thomson, 1889

... our eager-beaver researcher, undismayed by logic-of-science considerations and relying blissfully on the "exactitude" of modern statistical hypothesis-testing, has produced a long publication list and been promoted to a full professorship ... his true position is that of a potent-but-sterile intellectual rake, who leaves in his merry path a long train of ravished maidens but no viable scientific offspring.
—Paul Meehl, 1967

In almost every field, scientists often spoil their research with incomplete or improper data analysis. Even statistically sophisticated scientists make basic mistakes. By following five simple practices—inspecting the data, analyzing all of the data, preserving scales of measurement, estimating magnitudes, and documenting your work—you can avoid many problems and keep your research sound.

Inspecting the data

No matter how they are produced, data sets typically include errors of various types. One good way to detect some of these errors is to inspect the raw data directly. For most types of data, this means literally looking at all of the observations in a data set.

Inspecting the data is also a key task when collecting and managing data (see Chapter 6 and 7). Inspecting them again just before data analysis lets you catch errors that might have been missed in the earlier inspections, especially if you were not involved with those efforts. Reviewing the data also allows you to discover anomalies and gain insights that you might not otherwise, especially when you are in an analytic frame of mind.

Graphing (see Chapter 25) and/or tabulating distributions of variables is another essential part of analyzing data, even if you do not plan to include such graphs or summaries in your report. Statistical results sometimes can mislead if you do not understand the nature of the distributions. If your data set is so large that inspecting all of the raw data visually, observation by observation, is impractical, graphing and tabulating distributions may let you find some unusual observations that you can then investigate further.

Analyzing *all* of the data

Researchers can find different results depending on whether they analyze all of the data or just a subset of them. Analyze all of the data you planned to analyze, as indicated in your pre-registered study protocol (see Chapter 4).

There are a few legitimate exceptions to this principle. It is reasonable to exclude missing or invalid (e.g., out of range) data on key variables. However, do not exclude data because they have seemingly implausible (but possible) values. It is appropriate to exclude data due to measurement or methodological problems if you detected the problems *before* beginning data analysis. Researchers often discard data during data collection without even examining them when they notice a problem with the measurement process itself. If a measuring tool is prone to problems, test and calibrate it before collecting data for your main study. If you cannot resolve the problems, consider not using the tool in your work. But once you start collecting data with a tool for your planned research, there is no turning back: you have committed to analyzing all of the data you collect with it, no matter their defects.

Removing missing and invalid data from analysis, while necessary, can lead to bias in the results. Sophisticated methods for imputing missing data may be useful, but may also bias results. If you use such methods, report results both for when you exclude missing data and when you impute missing data.

Scientists vary in their opinions on removing statistical outliers from analysis. I believe it is best to analyze all of the data, including outliers. If you want to remove the outliers for some secondary analyses, specify your definition(s) of outliers in advance in your pre-registered protocol. Outliers often are genuine values, and excluding them can prevent discoveries and obscure patterns. Outliers, of course, may also represent errors in measurement and produce unreliable results.

Preserving scales of measurement

When collecting data, scientists typically strive to make the most careful and precise measurements possible. Yet their analyses often do not measure up, both literally and figuratively, with their data.

Scale of measurement refers to how the values of a variable can be compared. The four main scales of measurement form a hierarchy of increasing information content:

• nominal—values represent qualitatively different kinds, such as species, political parties, or chemical compounds;

• ordinal—values represent ordered categories, such rankings, geological strata, or social classes;

• interval—quantitative values fall on a continuum with no unique and non-arbitrary zero point, such as calendar date; and

• ratio—quantitative values fall on a continuum with a unique and non-arbitrary zero point, such as mass, age, and cell count.

Measurements on one scale usually can be transformed to measurements on a lower scale. For example, most variables that represent frequencies can be measured on a ratio (e.g., 0, 1, 2, 3, etc.), ordinal (e.g., none, some, many; 0–5, 6–10, 11–15, etc.), or nominal (present/absent) scales.

Researchers often degrade their data and research by converting variables to lower scales of measurement in analysis. This can mask patterns that exist, or create patterns that don't, in the original data. Lowering the scale of measurement for a variable throws away information and thus can lead to wrong conclusions. It is pointless and wasteful to focus on careful measurement when designing a study and collecting data only to reduce the data to cruder scales in analysis.

Classification is useful when studying natural categories or classification behavior. Typically, though, scientists use classification in data analysis as a way to simplify their thinking about a topic. That is, they tend to force nature into categories even when the phenomena are continuous. Also, researchers often categorize continuous data to permit analysis with their favored statistical techniques. This is an analytic version of the "law of the method" (see Chapter 50), in which researchers let their tools dictate what they study and how they study it.

Uphold the integrity of your data by preserving scales of measurement in data analysis.

Estimating magnitudes

Another way to measure up in data analysis is to use statistical *measures*, as opposed to statistical *tests*. The essential scientific goal in data analysis is to investigate questions such as "how much?" and "how strong?" Valid answers are estimates of magnitudes—descriptive statistics, whether univariate, bivariate, or multivariate. Researchers in many fields were trained to neglect statistical estimation and therefore often do not produce meaningful answers to their research questions. Regardless of your training, focus on estimating magnitudes and their associated uncertainties (e.g., standard errors or confidence intervals) to provide useful scientific results.

Documenting your work

Data analysis often involves many data, command, and/or output files, or a few such files each with lengthy content. In either case, use descriptive file names and document file contents with extensive notes or comments on the data files used, filters, procedures performed, analysis run dates and times, translations of results into natural language (e.g., "bimodal distribution", "X and Y strongly positively related", and "very similar outcomes for experimental and control groups"), and any other important details. Explain recoded, transformed, and newly constructed variables. Document in such a way that a scientist in your field who is not involved with the project could understand what your notes mean. This will help you or someone else understand your analyses and evaluate them as needed in the future. Likewise, delete files and content within files that you know are wrong due to mistakes you made; they are cancers that will confuse you when you later review your work.

* * *

So-called statistical hypothesis testing or significance testing is the prevailing mode of data analysis in many fields. The approach, regardless of the particular statistical technique, ultimately reduces to noting whether a probability value (p value) for a result is less than some arbitrary threshold level and therefore statistically significant. A p value is the probability the observed or more extreme result would occur if the null hypothesis (often representing no association, no effect, or a uniform distribution) were true. When random samples are drawn from a population, each sample tends to vary somewhat from the others simply due to chance. This is why results for a sample can deviate from the exact pattern specified in the null hypothesis when it is true for the population as a whole.

Even if you do not use or encounter statistical hypothesis testing in your research, the topic is nonetheless relevant to you. Statistical hypothesis testing pervades research in many fields with direct impact on nearly everyone's lives, such as medicine. Professionals, such as doctors, interpret evidence and advise others based on this approach. If only for your self-defense, read on.

Derivation of p values

A p value is simply a function of the sample size and the magnitude of association, effect, or departure from an expected distribution. Both of these quantities are already essential results to report. Sample size is an arbitrary factor determined by researchers or circumstances. A large enough sample size renders nearly any observed value of a statistical measure—no matter how small—as statistically significant, even for very low thresholds. Conversely, a small sample size renders many observed values of a statistical measure—even if large—as statistically not significant, even for high thresholds. Therefore, p values do not convey information solely about nature, but are influenced by how researchers conducted their studies.

Because a p value is a function of sample size and the value of a statistical measure, it can often be calculated from these results directly, especially when authors also report the precision of the estimate, such as a standard error or confidence interval. However, estimates of a magnitude of association, effect, or departure from an expected distribution usually cannot be derived (at least, not well) from just a p value and sample size.

Trivial questions and answers

In practical terms, statistical hypothesis testing involves evaluating analytic questions with yes or no answers, such as "is there a relationship?", "do groups X and Y differ?", and "are the observed data distributed non-randomly?" These questions are essentially meaningless from a scientific perspective. For example, in reality, nearly every variable has at least *some*, perhaps very small, non-zero relationship to almost every other variable. Rarely is a relationship exactly zero, or are data distributed precisely according to some chance expectation.

Sometimes researchers use statistical hypothesis testing to evaluate whether the observed data are consistent with an hypothesized non-zero, non-random value or range, perhaps derived from theory or previous research. In this situation, too, a yes-or-no result has little meaning. Is the hypothesis worthless if the p value is low? Is the hypothesis good if the p value is high? For these questions, statistical hypothesis testing is silent, because the sample size influences the outcome.

Significance testing is a sort of reverse alchemy. It involves statistical machinery to produce a non-numeric result—whether a null hypothesis is rejected—which is indeed a meager and unsatisfactory kind of knowledge. Both in science and life generally, the relevant questions concern not "whether" but "how much." We discern magnitudes and act on them—binary information alone is insufficient to guide our behavior or lead to understanding.

Making practical decisions

Decisions, though, do tend to be binary—choices to do or not do something. They are also inherently subjective. Yet many people want to escape the responsibility of making a decision and seek to offload the burden to an algorithm. Advocates for significance testing offer it as an algorithm for making practical decisions based on scientific research, regardless of its shortcomings as a scientific tool.

Fixed algorithms, such as significance testing, are problematic because they do not account for contexts and individual preferences. The crude, binary nature of significance testing compounds the difficulty. A quantitative estimate of a magnitude, however, gives far more information, particularly when paired with a measure of its uncertainty. This kind of input also allows decision makers freedom to put the result into their own individual contexts and factor in their preferences.

There is probably no more important a decision to make, based on scientific research, than whether to have a particular medical treatment for a life-threatening disease. If significance testing isn't relevant in this setting, then it's very unlikely to be relevant in making any sort of practical decision. Imagine the p value from a meta-analytic summary of a few, small, rigorous studies of a treatment is not significant, but the estimated effect is beneficial (and fairly uncertain). Would a patient—even one who is a trained scientist or statistician—be irrational to consider the treatment because the p value doesn't fall below some arbitrary threshold? Should a patient *ignore* an estimate of a beneficial effect that is fairly uncertain? I think the logical answer to both of these questions is "no." In fact, it might be irrational *not* to consider the treatment. Of course, patients often face a choice of multiple treatments and also weigh factors beyond effectiveness (such as side effects). But even for evaluating these, p values fail similarly as useful guides for a decision.

Statistical power

Statistical power analysis is a cousin of significance testing. It involves calculating a minimum sample size which would involve a reasonable chance of detecting a statistically significant result for a given value of a statistical measure. Researchers in some fields often use power analysis when planning the sample size for a future study. Significance testing is the ultimate purpose for power analysis. Therefore, the rationale for power analysis is weak.

Planning sample sizes may not be very important anyway. Limited opportunity, time, money, and/or effort often dictate the size of a study, and we simply accept whatever we can accomplish within these constraints. In general, I believe that many studies with small to moderate samples are more valuable than single or few studies with large samples. No single study is definitive, regardless of its size and quality. Scientific knowledge grows through replication, especially by independent teams of researchers (see Chapters 14 and 15). However, large, even very large, samples are especially useful for studies designed to produce public data sets for analysis by many scientists.

Temporary niche for statistical tests

Sometimes, there may be no good existing measure of association, effect, or departure from an expected distribution for a particular analytic problem. Statistical tests may be the only statistical tool available in such circumstances. I faced this situation for several years in one area of my research, and I had to rely on p values as my main analytic results. Eventually, I found a talented colleague (Giovanni Rinaldi) who overcame the formidable mathematical and computational challenges involved in the problem, and we developed an appropriate measure. If no suitable estimation methods exist for a statistical problem you have, strive to develop them.

Token compliance with empty rituals

You may be forced to include significance test results in your research reports to satisfy co-authors, editors, reviewers, and/or funders. Many scientists are firm believers in significance testing, or defer to those who are. As long as you report results for statistical measures and base your interpretations on them, there is no great harm in reporting significance test results.

9

Archiving and Sharing Data

Nullius in verba (on no one's word)
—motto of the Royal Society

Hypotheses come and go but data remain. Theories desert us, while data defend us. They are our true resources, our real estate, and our best pedigree. In the eternal shifting of things, only they will save us from the ravages of time and from the forgetfulness or injustice of men.
—Santiago Ramon y Cajal, 1916

A cardinal scientific practice is archiving or sharing the data that underlie study results. In some fields, publishing, archiving, or sharing of data has been the norm for decades or even centuries.

Data can comprise a variety of essential study materials. Sometimes the data include the ultimate objects of study, such as physical specimens. In other cases, the raw study materials themselves are perishable, not able to be preserved, or were never collected directly. Only the records of observations or measurements exist in these circumstances. An archive is public repository for these data, broadly defined. These include digital repositories as well as libraries of specimens, museum collections, tissue banks, or any other store of study materials. In most fields, archives were established only recently. In the past, many researchers might have been happy to archive their data had suitable archives existed.

Purposes and benefits

By making their data public, researchers demonstrate, at least nominally, that they did not fake their results. These data enable others to reproduce the original analyses to confirm the reported results or discover errors. The public data may also allow other researchers to investigate hypotheses and questions not examined by the original researchers. In some fields, scientists regard research for which the raw data are unavailable to other researchers as unscientific and bar such work from the literature. Researchers' reports are just words expressing claims. So to believe reports that lack underlying data is literally taking the author's word for it, contrary to the Royal Society maxim. Scientists who withhold data automatically make their research suspect.

Archiving your data also benefits you directly. In preparing your data for archiving, you organize them in such a way that make them readily interpretable to you in the future. Recently, I shared some data from a study I had completed 20 years earlier. The data and analyses were in several hundred files with file names restricted to eight characters plus an extension (a limitation of the operating system at the time). When I was collecting and analyzing the data, these names were sufficient for me to know exactly what they represented. But after two decades, it took some time to relearn what the abbreviations and codes meant. When I finished the study, I was aware of no suitable public data archive. So I closed the project without organizing all the material in such a way that would make sense to another person, much less myself in the distant future (even though I thought I would surely remember).

Archiving data also preserves them. Many, perhaps most, researchers do suffer some disaster—such as a flood, fire, equipment failure, or misplaced folder or box—that results in lost study materials. Thus, archiving data is like backing up data, but after the research is done (see Chapter 12). Data from countless studies have been lost forever because the scientists who collected them died without archiving them or making arrangements for their preservation.

Hollow objections

Some researchers complain speciously that it takes a long time to prepare data for archiving or sharing. On the contrary, the data should be ready for further analysis and not require extensive work to prepare if the researchers have already analyzed them. How else could they have obtained their results? If researchers haven't documented their data properly, archiving or sharing the data gives them a chance to do so.

Many scientists do not archive or share their data because they fear others may use the data in ways they dislike. For instance, others may perform analyses that they intend to do themselves, or conduct analyses that may produce results which differ from their original findings or are inconsistent with hypotheses they favor. However, researchers can choose to archive or share just those data necessary to replicate their reported analyses (although the ideal is to archive or share all of the data collected, analyzed or not). Researchers can also delay publication and data archiving until they have completed all analyses they plan to run on a particular data set.

Some researchers claim that the sensitive nature of their data precludes archiving or sharing them with other researchers. Many approaches to archiving and sharing sensitive data exist, nullifying this concern (see the last section of this chapter on "Sensitive data").

Other researchers do not archive or share their data so as to protect a commercial or military advantage. In such cases, their scientific publications serve primarily as advertisements, threats, or boasts.

Many journals and funding agencies have policies requiring authors and researchers to share data on request from others. However, journal editors and funding agencies may not enforce these policies. On several occasions, my colleagues and I have tried to obtain data from authors who had published their work in journals that had such policies but who refused to share their data. We appealed to the editors of the journals (including one of the leading general science journals) for help. In every case, the editors did not uphold their policies, nor did they sanction the authors or express concern about their published results. Public archiving is the only way to guarantee permanent access to data.

Archiving essentials

If your data set is small, include the data in your published research report, perhaps as an appendix. By doing so, you effectively combine archiving with publishing.

Many general and field-specific scientific data archives exist online for digital scientific data (see Notes). Most repositories for physical research materials have specific guidelines and standards for archiving. Digital archives, however, tend to be less prescriptive on the content and format of submissions. I recommend archiving the following when possible and appropriate:

• the *raw* data, as they existed prior to any adjustments or transformations you made;

• the final data used in analysis that do include any modifications you made;

• instruments (if in written form or expressed in programming code or pseudo-code), or detailed descriptions of them (for physical instruments; pictures and video are especially useful);

• data codebooks/dictionaries (see Chapter 7);

• details about methods and design not already published (see Chapter 6); and

• any other unpublished materials that verify what you did and help others understand your project.

The raw and final data often can be combined in a single data set. Some researchers also recommend archiving code for programs used to manage and/or analyze the data or the executable programs themselves. While certainly helpful, often this is not necessary or possible. Sometimes the code is proprietary. In most cases, the necessary procedures are standard and clear if the researchers described their methods and results well in their publications, appendices, and supplementary materials. When this is not so, it is critical to include the analytic software or code.

If possible, archive your data in generic, non-proprietary file formats that have a long history and are likely to persist into the future, such as comma separated values or other kinds of text files. Formats for proprietary software, even if they currently can be read by other software, are not likely to be readable as far into the future. Recently developed file formats, regardless of their merits or the corresponding software, are less likely to be used long into the future or be readable by a wide variety of researchers now. I have sometimes found an old data set, collected by another researcher, which I wanted to analyze but could not because the file format was obsolete. The corresponding software that could read it no longer existed on the computers available to me and no converter was available either, making the data unusable.

Even if you decide not to archive your data publicly, prepare them for your personal storage as if you were going to archive them. This preserves the data for future use by you or others, such as those who request them, including after you die.

Sharing unarchived data

Most scientific data are not in archives, and many will remain in the personal holdings of individual researchers. If we do not put data from prior research into a public archive, we are obligated to keep them ourselves in case other researchers ever request them. Naturally, we are obligated to share the data with *all requesters without restriction*. As long as sharing involves no new expenses for materials, equipment, or transportation (such as in shipping physical materials), it is unethical to ask requesters to pay anything for the data. Many researchers and institutions deny some requests for their data arbitrarily, maybe to prevent analyses that might contradict their own results and hypotheses. Results from such studies are therefore just as suspect as when researchers refuse to share the data with anyone.

Sensitive data

Sometimes, it is not ethical to archive or share *all* of the data for a study. In general, it is improper to include identifiers (such as names or addresses) of human research participants. Scientists often must ensure that individual persons or social units (such as families or communities) involved with research on sensitive topics cannot be identified even indirectly through combinations of other variables, such as geographic locations and demographics. Researchers usually must also exclude from archived and shared data the precise locations of vulnerable study sites and materials.

Several approaches to archiving and sharing data allow researchers to protect research participants, sites, and study materials in these and similar circumstances. One option is to remove the variables that could be used to identify, directly or indirectly, participants or sites from the archived/shared data set. Another strategy is to coarsen the data, such as by shifting to lower scales of measurement for some variables or adding random noise to parts of the data. A further approach is to create synthetic data sets based on models of the original data. If you use any of these strategies, repeat your original analyses with the modified data and include these results for reference in the archive or with the shared data. Yet another possibility is to allow others to analyze the data at a secure physical location with restrictions on analysis and data views to preserve confidentiality. The suitability of the different approaches varies across contexts, but at least one approach is possible for almost every study. And in the rare event none of these or other approaches work, you can analyze the original data on behalf of those requesting the data according to analysis plans they give you.

10

Using Secondary Data

In the mid-sixties Linton Freeman calmly announced to a gathering of busy researchers that there should be an immediate national moratorium on all data gathering. At the time he was interested in data banks and information retrieval systems, so the source of such a ridiculous notion was understandable, but of course none of the assembled ... could believe he was serious. Freeman was sensitive to the large amounts of unanalyzed and partially analyzed data on just about everything, gathering dust in files everywhere.
—Irwin Deutscher, Fred Pestello, and Frances Pestello, 1993

The masses of unanalyzed scientific data have only grown larger since Freeman made his bold appeal. There are several possible reasons for the accumulation of unanalyzed data. One is that the research grant funding system encourages projects involving data collection—with more staff, more time, and more money!— often at the expense of other research activities, including analysis. Overly ambitious scientists, regardless of funding source, also routinely collect more data than they will ever analyze (see Chapter 6). Moreover, many projects initiated by government agencies that involve large data collection efforts usually were designed by committees for general purposes rather than for particular research questions posed by individual, ego-involved scientists. This disconnect yields data sets that tend to be analyzed only partially.

Secondary data are data that others collected previously to investigate research questions that may or may not be the same as your current research questions. Secondary data may come from completed scientific studies or other sources. In designing and conducting their studies, scientists typically attend, at least somewhat, to measurement, sampling, and other methodological concerns. Secondary data from other sources, such as records kept by organizations and electronic devices or systems that automatically store information, generally reflect non-scientific goals and often do not conform to scientific methodological standards. Regardless of origin, secondary data can be particularly valuable when it is not possible or easy to collect similar data yourself. Even when it is possible to collect similar data, analyzing available data is efficient and economical.

The most important matter to evaluate when using secondary data is the extent to which the data correspond to your research question. Although some researchers get excited about large data sets simply because they are large, the amount of secondary data is irrelevant.

The greatest danger in using secondary data is that the availability and nature of the data determine the research question rather than the data merely serving as a context for examining a research question. Research projects (and even some sub-fields) developed in response to one or more data sets are prone to becoming empty scientific efforts because the logical order of scientific inquiry—question first, then data—has been reversed. This "data trap" is similar to the "law of the method," in which a researcher chooses the method before forming the research question (see Chapter 50).

A productive secondary data analysis project begins, like any other research project, with defining the research question(s). The next step is to establish the design requirements, including sampling and measurement, for a data set to meet before you consider analyzing it.

With those issues settled, then you can begin searching for suitable data sets. Public data archives are one natural source for data sets (see Chapter 9). Another source is authors themselves, when data underlying particular scientific publications have not been archived (see Chapter 40). Unfortunately, many authors ignore or decline requests to share data, despite their ethical obligation to do so. Regardless, such requests are worthwhile as sometimes authors do share. Secondary data that do not originate from scientific studies can be the most challenging to identify and access. Often, extensive searching, persistent requests, and sometimes, personal connections to gatekeepers, are necessary to find and obtain such data.

Before you begin working with a secondary data set, specify your analysis plans in a pre-registered protocol (see Chapter 4) to preserve the logical status of your work as hypothesis evaluation or description (see Chapter 51).

Secondary data analysts are at risk of misunderstanding the data—how they were collected, what they represent, and their deficiencies and limitations. Ordinarily, it is not good for the key components of a research project to be disconnected from each other. Secondary data analysis is, in some respects, a second choice for doing research. Ideally, the same persons are involved from conception to reporting, or if different sets of people are involved with different components of a study, they at least communicate well with each other so that everyone understands what the data are and their purpose. Even the most complete documentation may not cover everything of relevance. Similarly, secondary data analysts are also susceptible to bad data (full of error or bias) and fraud. If the data were fabricated or altered by the original researchers in some way, secondary data analysts might not be able to detect such corruptions.

Whenever possible, seek to learn as much as you can about the data by examining the source materials which underlie the data (specimens, event records, artifacts, or collection devices, for example) if they are still available. Interviewing the data collectors or those responsible for the data can give important insights, as can reading reports that describe, perhaps informally, the design and data collection procedures of the project. Sometimes such sources reveal critical details that influence your interpretation of the data and opinions about the most suitable analyses.

Cite and acknowledge the original researchers or organization responsible for the data in all reports based on your analyses (see Chapter 24). Send the final publication(s) to them as well to complete the scientific circle.

11

Programming

To me programming is more than an important practical art. It is also a gigantic undertaking in the foundations of knowledge.
—David Sayre, 1962

Those who would program must therefore be prepared to allow their thoughts to be subject to a relentless torrent of evaluation ....
—Donald Braben, 1994

Research in almost every scientific field now involves using computers. In some fields, programming is a universal, integral activity that all practicing researchers do. In most fields, however, some scientists are involved with programming while others are not.

Knowing how to program (write code) is an essential scientific skill, with both practical and intellectual benefits, even for those researchers who can't imagine any scientific need for programming. On the practical side, proficiency in programming allows you to do a variety of tasks for which no software exists. Most programming needs are only for small- to medium-sized tasks. Hiring someone else to program costs money you might not have, and also involves many inconveniences. If, as is common, the software doesn't work correctly or you need to modify it further, you will need the programmer again (unless he or she is part of your long-term research team). If the task is large and central to your project, progress in your research could be at the mercy of the programmer's availability, communication skills, and corresponding scientific knowledge, or lack of these. I find it is far better to be a mediocre programmer and fix my own problems for free on my own schedule, rather than be dependent on someone else who may or may not understand the task and handle it correctly.

Even with a basic knowledge of programming, you can have relatively informed conversations with programmers and technical professionals that you might not be able to have otherwise. This is especially helpful in many situations that arise in research:

- evaluating potential staff, students, and collaborators;
- communicating and working productively with professional programmers;
- assessing software to use;
- troubleshooting software; and
- appraising others' work involving programming.

Programming also provides something that scientific research does not: ultimate, provable solutions to problems. Science always entails uncertainty. Nothing can ever be known in science with absolute certainty. But in programming, you can work hard on a problem, and there is always a solution to it. You may need help from others to discover that solution, but whenever you break through and solve the problem, the joy you feel about having definitively solved the problem is unlikely ever to occur in scientific research (which, of course, involves other joys, just not about solving a problem beyond doubt).

The intellectual benefits of knowing how to program are as compelling as the practical benefits. Programming cultivates organized thinking and creative problem solving that are directly applicable to all aspects of scientific research. Consider these benefits of programming:

- gives exercise in making definitions and procedures explicit;
- forces you to adopt logically consistent thought processes;
- builds the habit of scrutinizing your work for errors and testing your tools for accuracy;
- requires you to conceptualize a problem both as a whole and as a set of interacting parts and sub-parts;
- reinforces the importance of documenting your work so that you can understand it in the future when you have

forgotten the details and are no longer immersed in the problem;

- improves your ability to organize processes and strategies efficiently;
- increases skill in quasi-programming, such as with spreadsheet formulas, macros in various computer applications, and code and commands for relational databases and statistical analysis software; and
- boosts your analytical and mechanical abilities.

In short, programming, like mathematics and statistics, opens new intellectual windows on the world.

If scientists in your field(s) tend to use a particular language, learn it. Otherwise, pick one of the several widely used programming languages, as they have the most extensive sources of potential help, such as online user communities, forums, tips, tutorials, and guides. These languages are also most likely to continue to be used in the future. General purpose programming languages have the widest applicability. Furthermore, languages that are relatively easy to learn and use are the best for scientists who want a quick payoff and the ability to write programs fast, especially for small tasks. Once you have learned one language, it is easier to learn a second language, because most languages share many of the same logical principles and approaches.

12

Backing up Research

One man was constantly employed in copying out all of our journals and scientific observations [from 2 years on a ship drifting in Arctic sea ice], etc., etc., on thin paper in a contracted form, as I wanted, by way of doubly assuring their preservation, to take a copy of them along with me [on a long trek by foot, dog sled, and kayak over the ice toward the North Pole and back to land].
—*Fridtjof Nansen, 1897*

A key mantra for survival in the wilderness is "three is two, two is one, one is none." That is, it is essential to have reserves of critical equipment and supplies, as well as alternatives in case of failure and depletion. If you have no reserves or alternatives, your life is at risk.

The same mantra applies to the survival of every scientific research project. All phases of research represent significant investments of labor, and possibly money and other resources as well. Each phase of research also may involve materials, experiences, and ideas that may be difficult or impossible to recreate if the records of the work are lost. Backing up research thus is a vital part of *doing* research.

Have at least three independent copies of data, analysis code and output (including images), and documents (including notebooks). "Independent" means here that if one storage location or access route were to fail, the other copies would remain intact and accessible. For instance, a digital file could be stored (and usually accessed) on a personal computer or networked server. Then, additional copies could be stored on an external hard drive and through an online service on a server at a different geographic location (the latter ideally far from the location of the other copies). Such an arrangement ensures survival of, and possibly continued immediate access to, your accumulated work in the case of equipment failure, fire, electrical power or Internet outage, natural disaster, or other mishaps. Of course, other storage media (including print) can be used depending on the circumstance and resources available. The key is to have independent, accessible copies of your crucial research information.

Backing up is especially important during data collection. If you cannot engineer automatic backup of data as they are collected, then schedule regular manual transfer from data collection devices or staff to minimize any potential data loss. Consider collecting duplicate specimens and storing them at separate locations. At the same time, you can also document events and physical specimens (with audio/video recording or other means) as alternate records of your source data.

If you want to preserve digital information over the long term —over many years or decades—have at least one copy that is not dependent on a particular technology, whether hardware or software. Technology sometimes fails, and can leave your backups unreadable and unusable. Technology also changes over time, often becoming incompatible with earlier versions, which renders your backups useless. Save at least one version of your computerized data, if possible, in a non-proprietary, simple file format, such as plain text. This helps ensure that your information will be readable long into the future, no matter how software changes (see Chapter 9). Printing on paper, of course, is a time-tested approach for preserving information for decades, although it can take much space to store a lot of information.

No one method of backing up is permanent. Physical and chemical forces cause decay in every storage medium over time. Therefore, to keep information for the long term, you will likely need to transfer it from an older storage device to a newer one, or even from an older technology to a newer technology. To stay ahead of entropy, the hand-offs in this relay race for most current digital technologies typically must occur every few to several years.

There is also a social component to backing up research. Make it so your research could be completed if you were to die or become handicapped unexpectedly. Give some trusted colleagues access to your materials and information, especially for projects in progress. This kind of insurance policy is usually built in to large collaborative projects, but it is easy to overlook for work involving few or no other researchers.

Being a scientist is a little like being a librarian. If you care about your information, then save it in multiple places and forms so that you and others can use it in the future.

13

Keep it Simple, Scientist (KISS)

Truth is ever to be found in simplicity, and not in the multiplicity and confusion of things.
—*Isaac Newton, 1643–1727*

Science should be so easy to do that a trained monkey could do it.
—*A. Kimball Romney, 1991*

The world is complex. Yet scientific understanding comes essentially through simplifying it as much as possible.

Simplicity makes your research easy for you to think about and track, which means that you can carry out your study effectively and with few mistakes. Simplicity also makes your work easy to communicate and understand, and thus likely that others will engage with the topic, perhaps even conducting their own research on it.

Beyond these general benefits, simplicity offers advantages in specific facets of research.

Research questions and designs

Honing a research question to its bare essentials gives clarity of purpose. A simple research question enables you to develop the most appropriate and consistent design, methods, statistics, and interpretations of results for a study. Limiting the empirical scope of a study helps to prevent complex research questions.

Simple research designs pave the way for simple statistical analyses and interpretations of results. Elaborate designs, in contrast, introduce complexity to the rest of the project and are less likely to be replicated by others. Developing simple research questions and designs usually involves some time, effort, and discipline as part of a careful research planning process (see Chapter 3).

Methods and statistics

Simple data collection methods are the easiest to teach others and thus enhance replicability (see Chapter 14). Easy-to-use methods of all kinds may be less likely to produce errors. Sophisticated statistical techniques do not necessarily signal rigorous research. In general, scientists use advanced analytic techniques as partial compensation for complicated, problematic, or weak research designs. Statistical methods can never reverse defects in design, regardless of researchers' wishes to the contrary. Complicated statistical analyses with many variables, interactions, adjustments, and controls also tend not to be replicated.

Each additional layer of data transformation, processing, and analysis takes an investigation to a more abstract plane. Some or all such steps may be necessary, but they come at a price. Results become less intuitive, since, to understand them properly, it requires mentally repeating the steps and keeping them in memory. Each step also introduces the possibility of error and allows researchers to draw out or select favorable results and hide less desirable results. Furthermore, each additional step makes it more difficult to scrutinize the overall effort in terms of logic and accuracy of execution.

Scientists sometimes use complexity in their data analysis to intimidate other scientists. We researchers like to impress others with our technical sophistication. Researchers less expert on specific methods who question the results from complicated analyses risk being dismissed as ignorant. Most who have such doubts do not question the research publicly so as to avoid embarrassment. This dynamic undermines the skepticism essential to building reliable scientific knowledge (see Chapter 61).

Theories, models, and explanations

Simple statements, if knowledge is our object, are to be prized more highly than less simple ones because they tell us more; because their

empirical content is greater; and because they are better testable.
—*Karl Popper, 1935*

Complex theories, models, and explanations with many elements and relationships are traps. If any component is invalidated empirically, what does that imply for the fate of the whole? If it doesn't mean that the theory, model, or explanation should be rejected, what are the objective criteria for assessing its empirical fit? Simple hypotheses and theories speed scientific progress because they tend to be evaluated, and discarded, more quickly and easily than complex ones.

Complex models, theories, and explanations also fail in other ways. If the world is complex and a theory or explanation is also complex, what is the value of that theory or explanation? How has it increased understanding?

Complicated explanations do not reflect sophistication. Rather, they represent desperation and confusion.

Great scientists and thinkers are not those who are so brilliant that few understand them, but whom many understand due to the clarity, logic, and simplicity of their thought in relation to the evidence.

Although we scientists try to reduce the complexity of the world to comprehensible principles, we would be wise to heed Alfred North Whitehead's caution: "We are apt to fall into the error of thinking that the facts are simple because simplicity is the goal of our quest. The guiding motto in the life of every natural philosopher should be, 'Seek simplicity and distrust it.'"

Projects

Research projects themselves can become quite complex, even independent from the complexity of the research itself. The process of competing for grant funding (and thus salaries) often leads researchers to propose (and do) large and complex studies in hopes of impressing reviewers and funding agency staff. However, usually smaller and simpler research projects would be better, with longer gaps between projects so that the researchers could consider results, implications, and future plans more carefully. The grant and contract treadmill often promotes expensive, unnecessary, wrong-headed research.

A perennial problem we scientists (and people generally) have is attempting projects that are not feasible. We underestimate how long, costly, and complicated projects will be and are overly optimistic about factors outside of our control (see Chapter 3). Modest aims and unambitious schedules make for realistic and successful projects. In research, it actually is better to attempt to do too little than too much. If a project is not completed, there might no useful results, despite months or even years of work.

Large projects involving many staff become bureaucracies (see Chapter 35). While this formal division of labor and communication structure may be inevitable given the size of the projects, such organization carries risk. Error is much more likely when the lead scientist(s) cannot do or explain in detail each task in a study. Formal organization also tends to crush creativity and independent thinking and slow progress generally. Ideally, it may be best to have just a handful of researchers of any level involved in a project, and even fewer project leaders.

Project coordination also can become unwieldy with many study sites, supporting vendors, or staff roles. Moreover, complicated study schedules, with many different activities occurring concurrently and starting and stopping at different times, can become logistical nightmares and jeopardize project success. Projects with many moving parts tend to have frequent breakdowns and rarely provide scientific value commensurate with the extra effort they require.

Avoid other bureaucratic snarls to the extent possible. Multiple funders for the same project can complicate work with their different demands (including reporting requirements). In research on people done in university and other institutional settings, ethics or human subjects review committees may hamper, alter, and delay studies. In this review process, scrutiny of proposed research tends to be inversely proportional to risk:

committees may dictate damaging changes to study protocols to prevent the slightest imagined potential harms, yet paradoxically approve unnecessarily dangerous or deceptive procedures in other projects that seemingly offer little benefit. Avoid such interference by designing research, whenever possible, that is exempt from ethical review or involves the fewest elements likely to trigger censors.

Assessing complexity

One way to assess how simple or complex your planned research is to write your plans in a study protocol (see Chapter 4) that includes the research questions and full logistical details. If you have difficulty writing the protocol clearly, that is a first sign your plan or some part of the study is too complex. For a stronger test, ask colleagues to review your written plans, focusing on complexity (see Chapter 31). Their feedback is critical, especially if you have limited research experience or are attempting a different kind or scope of work than you have previously. Over the long-term, as you do research and think about ideas and future research plans, learn from past projects that suffered from complexity and look continually for ways to simplify your work.

14

Replicating Research

... particular facts are never scientific; only generalization can establish science.
—*Claude Bernard, 1865*

The smallest group of facts, if properly classified and logically dealt with, will form a stone which has its proper place in the great building of knowledge, wholly independent of the individual workman who has shaped it. Even when two men work unwittingly at the same stone they will but modify and correct each other's angles.
—*Karl Pearson, 1892*

Replication is the bedrock of science. No single study is ever definitive, no matter how large it is, how rigorously it is done, or who does it. Individual studies, or more accurately, particular results from them, are themselves data points. How are results related to the time and place of a study? The particular researchers who conducted a study? The specific methods, materials, equipment, measures, and samples they used? Only in summarizing the data points that represent different studies can valid assessment of a research question begin (see Chapter 15).

Any study worth doing once is worth doing twice. And any study done only once is not much better than it not having been done at all. Without replication, a single study has fairly little value. This is not a matter of statistics, but of logic and sociology. A single study is inherently untrustworthy because it was done only once. Every study is susceptible to error, bias, and fraud. There currently is no foolproof way to demonstrate they are entirely absent from a study, although the essential practices I highlight in this book and specific procedures in particular fields can build confidence. Reliable knowledge, by definition, comes only from different researchers with different ideas working on the same problem in the same (or similar) ways in different places and times. And even when these conditions hold, the resulting knowledge is *still* provisional.

Results from a non-replicable study are effectively an anecdote. Unfortunately, such research characterizes a large proportion of work in some fields. This is especially common in research on one-off programs and other interventions implemented in real-world settings, whether in human affairs, the natural world, or the technological arena. Such programs are often unique or have unique combinations of components, making them incomparable with other interventions focused on the same or similar problems. Consequently, systematic summaries of such efforts are not possible, thus stunting reliable scientific knowledge.

Replication refers to repeated similar *studies.* Replications might produce different *results.* In systematic reviews and meta-analyses, researchers address the consistency of results across similar studies (see Chapter 15). Repeating an author's analyses with the author's data to confirm the author's reported results also is not replication in the broad scientific sense.

Plan your study design and analyses with future replication in mind. Make it so that your study or parts of it can be repeated, by you or others. Use measures, materials, equipment, and methods that are readily available and others can use easily. Researchers often choose statistical analyses that make their reported results non-replicable. For example, for studies that involve many variables, other scientists may not always be able to study the same set as you did. If you report results for multivariate analyses only, your results may become non-replicable. Therefore, in such situations, report univariate and bivariate results comprehensively as a descriptive resource, perhaps as an appendix (see Chapter 20). Even better, archive your data so that other scientists can more easily replicate analyses with the variables they have in common with your work (see Chapter 9).

Planning research that replicates others' work as closely and wholly as possible is a useful pursuit. But partial replications, in which you repeat some elements of prior research, are also worthwhile. This allows you to do novel and repeated research in the same study. Replications also do not have to be identical to the original studies. By changing details of the context and procedures in repeated studies, you can assess the generality of the phenomenon under study.

In some fields, researchers prize novelty in research so much that they denigrate replication or consider it less valuable than other research. In these fields, researchers may rarely replicate their own or others' research. If you are working in such a field, then it is your responsibility to replicate your own research. Otherwise, you may never know how reliable your results are. In fact, whenever possible, plan to replicate each new study you do, perhaps with minor changes to procedure, no matter what the results of the first study show or how common replication is in your field.

How much replication is enough? The answer may vary from problem to problem. One indicator that no further replication may be needed is low variability in results despite great diversity in the studies in terms of researchers, their hypotheses and ideologies, methodological and design factors, and sponsors. Further replications may also be unnecessary if there is some degree of variability in results across studies, but scientists can account for that variation empirically and reasonably. Of course, these judgments are fairly subjective, but no replication study is worthless.

15

Reviewing Systematically

There are times when I feel compelled to view our knowledge as a large boutique or department store that is not well organized and has no inventory, because we do not ourselves know what we already have and hence cannot make it serve our needs. There are infinite numbers of worthy thoughts and useful observations to be found in published works ... and if ... these were to be collected and arranged by topic ... we would genuinely appreciate our rich heritage, and would ascribe our blindness to not having taken proper advantage of these worthy thoughts.
—Gottfried Wilhelm Leibniz, 1688–1690

Individual studies are not the endpoints of scientific research. The result from an individual study is just one piece of evidence —a single observation—amidst the results from other studies, if they exist. Scientific understanding comes only when researchers review the literature on a topic systematically by finding and weighing such pieces of evidence from different studies in a rigorous way. Systematic reviews are the ultimate empirical statements on a topic.

One reason that systematic reviews are powerful is that they involve searching the literature thoroughly, when done well (see Chapter 5). Neglected or even previously uncited works often have crucial results. Researchers' knowledge of the literature on a topic is almost always incomplete and biased, so formal search procedures are necessary.

Objective summaries of evidence involve some sort of quantitative synthesis. In many fields, this approach is called meta-analysis. The synthesis usually includes a summary estimate of the quantity or quality under study as well as indicators of the precision of the estimate and the amount, consistency, and quality of the underlying evidence. To produce a synthesis, there must be multiple studies that are similar enough on the same research question that their results can be compared and combined (see Chapter 14).

A systematic review is not always a summary of prior research on established questions. A systematic review may cover questions that previous researchers did not investigate directly in the studies they reported. A systematic reviewer may take the evidence from their reports, or their underlying data, to examine these other questions.

Systematic reviews are studies of studies. Therefore, all of the principles and practices for good primary scientific research apply to systematic reviews, including framing the problem accurately (see Chapter 3), pre-registering study protocols (see Chapter 4), emphasizing estimation rather than significance testing (see Chapter 8), interpreting results in a balanced way (see Chapters 21, 58, and 61), and recognizing the limits of the research designs of the summarized studies (see Chapter 52). Protocols for systematic reviews describe the literature databases and resources to be searched, how they will be searched (such as keywords and forward and backward citation searches), explicit criteria for including and excluding studies, information to be extracted from included studies, and methods of analysis, among other details.

Just like any other type of research, systematic reviews vary in quality. The validity of the results in a systematic review depends on several factors: the relevance of the evidence to the research question; completeness of the literature search; appropriateness of the inclusion and exclusion criteria; suitability of the definitions, designs, and measures; and soundness of the analytic techniques. Moreover, if the literature on a topic is biased, any summary will be biased, too. Current conventional publishing practices in science tend to skew the published record (see Chapter 32), and the degree of distortion may vary by field, topic, and country.

Unsystematic reviews

Although systematic reviews are now common in many fields, unsystematic reviews abound. Narrative literature reviews, textbooks, commentaries, introduction and discussion sections in research reports, and book length overviews of scientific topics all tend to involve selective reviews. They have a free-form structure, typically following the direction of the author's argument. There is usually no way to know what evidence the author found but did not include, or even how the author went about defining or finding the evidence. Has the author highlighted work in line with his or her ideas and ignored contrary results? Countless times I have trusted an author's informal summary of the evidence only to find later, in my own independent search of the literature on the same topic, a very different picture.

Unsystematic reviews are not scientific summaries. They are, in effect, stories. We humans are naturally inclined to tell and receive stories. Stories have a place in science, especially in illustrating how to do research, communicating complex ideas, critiquing research, making philosophical or methodological points, and motivating new research. Stories can also convey the findings from systematic reviews, but even in this case, the resulting picture can be distorted by what the author chooses to include and exclude.

Consensus statements, practice guidelines, and standards

Historically, the claim of consensus has been the first refuge of scoundrels; it is a way to avoid debate by claiming that the matter is already settled. Whenever you hear the consensus of scientists agrees on something or other, reach for your wallet, because you're being had.

Let's be clear: the work of science has nothing whatever to do with consensus. Consensus is the business of politics. Science, on the contrary, requires only one investigator who happens to be right, which means that he or she has results that are verifiable by reference to the real world. In science consensus is irrelevant. What is relevant is reproducible results. The greatest scientists in history are great precisely because they broke with the consensus.

There is no such thing as consensus science. If it's consensus, it isn't science. If it's science, it isn't consensus. Period … Consensus is invoked only in situations where the science is not solid enough."
—*Michael Crichton, 2003*

Reports of consensus statements, practice guidelines, or practice standards are not systematic reviews and are not scientific summaries of evidence. Rather, they are political documents, in at least three senses. First, the sponsors of such efforts are usually politically-involved organizations, such as governments, industry groups, and professional societies. Second, authors and sponsors develop these documents in a political way. They pick which scientists participate, inevitably excluding those with unconventional views opposed by other participants or sponsors. Social dynamics and the varying political influence of participants unavoidably shape the deliberations and conclusions of the collective effort. Sometimes, participants conduct or consult systematic reviews during their deliberations. That can improve the quality of their deliberations, depending on the quality of the systematic reviews, yet their conclusions still are ultimately subjectively and politically determined. Finally, these documents serve political, not scientific, purposes. The target audiences for these documents are government officials, clinicians, other professionals, or the general public, and the authors seek to influence decisions and behavior outside of the scientific realm.

16

Doing Research without Funding

'Endow scientific research and we shall know the truth, when and where it is possible to ascertain it;' but the counterblast is at hand: 'To endow research is merely to encourage the research for endowment; the true man of science will not be held back by poverty, and if science is of use to us, it will pay for itself.'
—Karl Pearson, 1892

We had no money, no suitable laboratory, no personal help for our great and difficult undertaking. It was like creating something out of nothing ....
—Marie Curie, 1923

We have no money to spend, so we shall have to think.
—Ernest Rutherford, early 1930s

... funding is by no means necessary for creativity, in fact the relationship may be inverse.
—Philip Anderson, 2011

No matter your employment situation, you can always do scientific research; funding is not required. Indeed, the world is open to observation and experimentation by anyone, regardless of formal training, experience, or station in life. Most scientific research before World War II in most fields was not funded specifically. Since then much, if not most, research has been financed by governments, universities, industry, private foundations, and other organizations with funds dedicated to particular projects.

Despite this, unfunded research (or self-funded research when there are small costs), plays a critical role in science. Unfunded research is typically free of bureaucratic hassles and less likely to involve financial conflicts of interest. Funded research can be compromised by interference from those not doing the actual work, such as bosses, clients, and funding agency officials. Scientists doing unfunded research can preserve the integrity of their work and avoid conforming to scientifically empty fashion, political correctness, and sponsor biases.

Unfunded research also tends to focus on the most relevant research questions and involve efficient study design. Unfunded or self-funded researchers bear all the risk and are reluctant to waste their unpaid effort (and sometimes money) on frivolous or sloppy research. Ample funding can actually hinder scientific advances in part because scientists are rewarded even when their research is irrelevant or inefficient. Labors of love may be the most valuable scientific projects of all.

Doing research without funding is not a requirement to be a good scientist, but good science is often unfunded or self-funded. Moreover, without significant unfunded research, areas of study are prone to slow progress and unreliable knowledge.

Some researchers regard unfunded research as low status, given that it hasn't been officially approved in advance by other scientists and institutional authorities. In their opinion, it is unseemly to perform research not financed by a government agency, corporation, or foundation. Such views, however, discourage research that may challenge the status quo or address topics that have been ignored. Ultimately, prejudice against unfunded research is about restraining other scientists' freedom and maintaining externally funded researchers' status.

For scientists who are not tenured or tenure-track academics, unfunded research also signals commitment to science rather than simply to income-producing careers. Is scientific research just a way to make a living or is earning a living a way to do scientific research? Artists, musicians, writers, filmmakers, inventors, entrepreneurs, and others frequently pursue long and labor-intensive projects with no outside funding. Scientists can do the same if they are committed to building knowledge. In fact, so-called amateur scientists (those without full formal scientific training) make important contributions in many fields, such as astronomy and paleontology, without pay or support. Such dedication represents the true scientific spirit.

There are many options for doing unfunded research. If you are employed as a scientist, you may be able to use facilities, other resources, or your position itself to conduct unfunded research on your own time (or even on your employer's time, such as in academia). If you are employed in an applied practice setting, you may be able to do research as part of your regular work without funding if the research addresses a matter relevant to your job. If you aren't employed by an organization that has a good platform for doing research, consider whether colleagues might be willing to offer theirs, perhaps as part of a collaboration.

Good research can still be done even without job-related access (your own or colleagues') to facilities and resources. It may be easier and less expensive to do unfunded scientific research in a completely independent fashion now than at any point in history. For computing, there are free, robust operating systems used extensively by scientists (such as Linux) and excellent free software for general purpose scientific work (such as word processors, text editors, reference managers, relational databases, programming languages, spreadsheets, and statistical analysis packages) as well as many specialty tasks on all major operating systems and online. The Internet gives researchers free access to countless data sets, large portions of the scientific literature, potential research helpers (citizen scientists who collect, code, or analyze data), potential collaborators, and communities of fellow scientists (such as in online forums and e-mail lists). In many cities, there are community wet lab spaces for biological research available for no or low cost, allowing researchers access to shared equipment and materials. Resourceful researchers can sometimes even develop substitutes for expensive technology. A beautiful example is C. V. Dharmadhikari's construction of highly productive scanning tunneling microscopes from junk and very inexpensive materials in India beginning in the 1980s. Unfunded researchers' will to learn can spark creativity that otherwise would lie dormant with a healthy research budget. To do unfunded research, all a scientist truly needs is desire and some time.

Of course, there may be lines of research that are difficult or impossible to pursue even with creativity, determination, and free or low cost resources. Scale the size, scope, and nature of your research to your available time and opportunities.

There are several kinds of research that can always be done without funding, including systematic reviews/meta-analyses (see Chapter 15); studies with equipment and materials you own, can make, or buy for little money; simulations (see Chapter 54); development of statistical and mathematical methods; development of theory; secondary data analysis (see Chapter 10); analysis of previously collected specimens and materials (such as at museums); and critiques of prior research. It is perfectly fine to study small problems that do not receive much attention. The key is to find something that interests you, even if you are not a specialist on the topic (see Chapter 49).

17

Checking—Part I

When we look closely, we recognize the same balls being dropped over and over, even by those of great ability and determination. We know the patterns. We see the costs … Try a checklist.
—*Atul Gawande, 2009*

[Professionals who use checklists] improve their outcomes with no increase in skill.
—*anonymous director of a multibillion dollar investment fund, 2009*

I wrote this book as a reference for myself and other scientists. All researchers overlook some of the essentials at least some of the time. Few of the essentials are covered in textbooks or courses, so they are not standardized or institutionalized practices. They require individual initiative to apply. Many of the essentials go against our selfish, emotional natures. Using this book as a checklist helps instill the discipline to implement the essentials and overcome inertia.

The consistent use of checklists has dramatically improved safety and efficiency in varied human pursuits, such as aviation, large-scale construction, investing, and medicine. Some methods in science involve checklists or checklist-like procedures. There are also checklists for the necessary information to include in research reports for some fields (see Chapter 30). Both kinds of checklists are useful and important, but focus on very small parts of the overall research sequence. We can advance reliable scientific knowledge by using checklists for all steps of the research process.

The Appendix contains several checklists. The two main ones focus on pre-study activities (from conception through preparation) and writing a research report. I have also included lists for other activities that occur throughout or independent of particular research projects.

Each checklist is simply a list of chapters from this book, to be read and considered again. Most of the checklists include chapters from each of the four sections of this book. For the pre-study and writing checklists, the chapters can serve as prompts for you to do the corresponding practices or confirm that you have done them. The pre-study and writing phases of research occur over days to years. Hence, you can take time to review the chapters and reflect on their implications. You can return to chapters on a checklist repeatedly as situations or your ideas change. These checklists naturally tend to focus on abstract rather than concrete activities, in contrast to checklists in most other professions. As a result, these checklists require your prolonged effort and motivation for effective use.

Some professionals oppose checklists as constraints on their work. The checklists in this book do not limit scientists in any way. Even with the checklists, researchers still can use their creativity, insight, and skills without losing any scientific freedom. On the contrary, the checklists remind researchers to *exercise* their scientific freedom.

The pre-study checklist covers some activities that occur after you have started a study. This is because doing a study usually means simply carrying out study plans. The time to consider those activities is when you prepare the study. Of course, sometimes researchers need to change plans once a study is underway due to circumstances. When this happens, consult the relevant chapters from the pre-study checklist again. A pre-registered protocol (see Chapter 4) is an excellent project-specific checklist for executing a study plan. Likewise, the checklists in this book are not replacements for, but critical supplements to, any checklist you might already have for one or another step in your research.

Research projects involving multiple researchers often have a hierarchy and a division of labor, with different persons working on different tasks. Nonetheless, the checklist process, especially for the pre-study phase, improves to the extent that more project researchers are involved. Open discussions in which all researchers participate actively are ideal. Such discussions are likely to be productive if project leaders not only tolerate but welcome criticism and questioning from others that may occur in the process.

Checklists may make your work more efficient. Using checklists helps to discard bad ideas early and prevent wasting time on them. Similarly, you can discover and prevent problems before they stall or imperil a project.

Most problems in scientific research can be avoided by following the essentials. Following them in a structured way is probably the best approach to ensuring you apply them.

Communicating Research

The first several chapters in this section focus on the common parts of a research report: Introduction, Methods, Results, and Discussion. The specific structure of a research report tends to vary across fields and contexts. Sometimes, these sections go by different names, or may be merged. Their content can vary as well. Regardless, my advice applies to virtually all kinds of research reports, as these sections serve basic functions for scientific communication.

18

Simplifying the Writing Process

True ease in writing comes from art, not chance,
As those move easiest who have learn'd to dance.
—Alexander Pope, 1711

Do whatever is easiest first.
—Howard S. Becker, 1986

Writing a research report or even an abstract can seem like a daunting task. It involves more information that can be manipulated in our minds at once. Many scientists, and people generally, make such tasks possible, even easy, by breaking them into smaller tasks that can be accomplished readily.

Scaffolding

The approach I recommend requires no pondering important things to say and no inspiration for creating material. There should be no such thing as "writer's block" for scientists. With this approach, there is always a scaffold that exists before you even start to draft any section of the report. The scaffold for the Methods section is your pre-registered protocol (see Chapter 4) and documentation you have made during data collection, management, and analysis. The scaffold for the Results section is your results themselves. The scaffold for the Discussion is your Results section, prior work you identified in your literature search (see Chapter 5), and notes and documentation you have made during your research project (see Chapter 46). The scaffold for the Introduction is the rest of your report along with your pre-registered protocol, prior work you identified in your literature search, and notes you have made during your research project. Finally, the scaffold for the abstract is the full draft of the report.

There is a natural sequence in writing the sections of a research report. Begin with the Methods section (see Chapter 19), starting after you have set the design and collected the data (or even while you are still collecting data). If you can, write the Methods section before analyzing the data. This will help prevent your reporting what you *did* from being influenced by what you *found*. The methods do not change based on the results.

Next, write the Results section (see Chapter 20). Write it in stages, drafting text and creating tables and figures for sets of results as you obtain them. This way, the results are fresh in your mind when you report them, and subsequent results will not affect how you describe earlier results.

Then, write the Discussion section (see Chapter 21). By writing it after the Results, you keep your interpretations and speculations, at least those committed to a draft, from influencing what findings you report and how you report them.

Finally, write the Introduction (see Chapter 22). You cannot introduce something until you know what it is, or give directions to a destination without knowing the way. Similarly, it is not possible to introduce a research report fully or correctly until you have written the rest of it (or at least outlined it in fine detail). With all of these sections in place, then you can write the abstract with ease.

Notes are a very valuable resource in writing a paper. For every paper you write, keep a document with notes and ideas for the paper, categorized into the separate sections of Introduction, Methods, Results, and Discussion. At any point in the research process—even soon after conceiving the idea for the work—store in this document all the relevant thoughts you have about the study and possible points or elements to include in the paper. These thoughts do not have to be organized or even consistent, because you can sort them out later. The key is to capture everything in writing that might possibly be relevant. Memory is fragile for everyone, and it gets more fragile as we get busier and older.

If you do not have a protocol, study documentation, or notes, I suggest creating this content retrospectively before starting to write the paper. To do so, brainstorm or free write all ideas and information in whatever way, order, or form to get the missing raw materials. This lets you separate *creating* content from *communicating* content.

If you do research in a structured way—planning your study carefully, pre-registering your procedures, and documenting your work and making notes throughout—then using the approach to writing I have described will prevent you from getting stuck on what to write. It will also help ensure that your report is accurate and complete.

Pile sorting

Pile sorting is a research method for judging the similarities and differences among a set of concepts or physical objects. The same basic technique can be used in writing a research report.

The process starts with some unorganized or partially organized material, such as notes and documentation corresponding to a specific section of your report. The material might consist of whole paragraphs, single sentences, sentence fragments, and/or even single words that might represent larger points or ideas. In pile sorting, you take these separate pieces or chunks of text, and put like with like. In an electronic document, you create "piles" in your draft by putting similar chunks together on the same part of a page or separate pages, making sure there is space between separate piles. (Pile sorting also works well when writing on paper.) If a chunk belongs in more than one pile, you can copy it into each pile where it belongs. Piles can be as big or small as the material warrants. Group easy chunks first. If some chunks are difficult to classify, postpone your decision until you have sorted all of the easy ones. By then, it may become apparent which piles suit the hard chunks of text best, or perhaps you will see a need for new piles to accommodate them.

As you sort chunks of text, consider labeling each pile with one or a few words that indicate its purpose or content, as these themes occur to you. This makes sorting easier and faster. These labels can eventually serve as the kernels for sub-section headings or topic sentences of paragraphs.

After you have sorted all the chunks of text into piles, you can then split or merge piles as necessary. If a pile has many ideas, points, or sentences, divide it into smaller piles of highly similar content, possibly including new labels for the smaller piles. If a pile has a single idea, point, or sentence only, try to merge it with another pile to which it is most similar, or make a new pile of miscellaneous chunks. In splitting and merging piles, aim to create piles that correspond roughly to paragraphs in terms of amount and similarity of content. You don't need to write paragraphs at this point; you are just putting together the elements of paragraphs.

During this phase, also look for piles to eliminate entirely if their content is tangential, contradicts other material, or is otherwise weak. Create a "parts cut" file paired with your draft where you keep the content you remove, so that you can retrieve it later if you change your mind (see Chapter 29).

Next, arrange the piles into the order in which you want them to appear in your draft. If appropriate, put sets of adjacent and related piles under sub-section headings.

In the final step, arrange the chunks of text within a pile into the order which you want them to appear in the resulting paragraph. Identify similar points or ideas within the pile and remove redundancies. If two points or ideas are similar but not the same, merge them, or pick the stronger one and discard the other. My accumulated notes for a study often have many redundant or similar ideas.

Now this section of your draft is well-organized in terms of content, and all that remains is to compose complete sentences and full paragraphs, edit thoroughly and repeatedly, and proofread (see Chapter 30). Of course, all throughout the pile sorting and ordering process, you can make adjustments as you see fit. You can even compose whole sentences or paragraphs within piles if it is easy.

Managing references

Use a reference manager to organize your sources, cite them in the text of your reports as you write, and build bibliographies quickly and efficiently. Many free programs exist. They save a lot of time whenever you need to add, edit, delete, or reformat the references and citations in a manuscript.

* * *

Scientists vary in how they go about writing a scientific report. The approaches I have described are just a few of the alternatives. Nonetheless, these approaches are likely to make the writing process efficient, effective, and even fun.

19

Writing a Methods Section

I keep six honest serving-men:
(They taught me all I knew)
Their names are What and Where and When
And How and Why and Who.
—Rudyard Kipling, 1902

Scientists try to understand the world empirically. Through observation and experiment, we seek to build reliable knowledge.

In our scientific articles and other reports, we describe what we *did* and what we *found*. Usually in our reports we also discuss hypotheses, theories, interpretations of results, implications, and/or speculation. But these are all secondary matters. The most important parts of any scientific report are the Methods and Results sections because they contain the evidence. In some fields, the Methods and Results appear in sections with other labels that are sometimes unique to particular reports. Regardless, the methodological information I highlight in this chapter should be included somewhere in the report.

The information in the Methods section (or "Materials and Methods" or "Method" section) allows readers to make sense of and evaluate the results. A Results section without a preceding Methods section is like a punch line for a joke without the setup. Unfortunately, some journals now put the Methods section at the end of an article, in small print, or remove it from the main article entirely, which undermines scientific understanding and de-emphasizes the heart of research.

A good Methods section includes a full enough description of study procedures to enable other researchers to replicate the research fully (see Chapter 14). This is crucial. Why should anyone care about a study that can't be repeated?

Specifically, in the Methods section, authors answer the 5 Ws and one H. These are the questions that Kipling wisely recognized as the keys to knowledge: who, what, where, when, why, and how.

The Methods section is the easiest part of a scientific report to write. We already know what we did, and no interpretation is required. In fact, a pre-registered protocol for the study (see Chapter 4) and notes on progress and relevant events during the research (see Chapter 6) give all the information needed to write a Methods section. Material from these sources might not even require much editing. Still, sometimes authors leave out essential information. By checking whether we address the fundamental questions, we can reduce the chance of omitting something important.

Who?

Many authors write Methods sections, often in passive voice, as if no researchers were involved in the study (e.g., "specimens were collected", "surveys were administered", "samples were taken"). Instead, use active voice to identify who carried out particular tasks in a study (see Chapter 27). Often it is sufficient to use "I," "we," or the name of a third party as the subject in such descriptions. Sometimes it is useful to name specific authors or other individuals when a task requires special expertise.

Note when the sample of specimens or participants is the same as or overlaps with a sample from another report (or study). More generally, if the study is the same as that in another report (but focused perhaps on a different outcome or topic), cite the other report(s). Otherwise, readers may believe, incorrectly, that the separate reports represent different studies.

What?

The Methods section covers study design and procedures as well as the equipment, facilities, materials, and software researchers used to carry them out. Despite this, authors sometimes neglect relevant details, such as the sources (including manufacturer and city) of the equipment, materials, or software. For the most important methodological aspects of a study, be very explicit. For instance, if a study involved randomization (e.g., random assignment or probability sampling), describe the source of the random sequence (computerized random number generator, random number table, digits derived from atmospheric noise, etc.).

Where?

At least for field studies of any kind or research involving people, report the study location (at least country/region of country, and ideally city or town as well). In many cases, the specific geographic, ecological, and/or social context is also pertinent. Even when you report a laboratory study, include information on any aspect of the setting that could possibly influence your results. This means that the name and location of the laboratory in which you conducted the study is usually relevant, although many authors fail to report this information.

When?

In Methods sections, authors often omit the date of data collection or other key study activities. Present a more complete picture by reporting at least the month(s) and year(s). Sometimes, specific days, days of the week, or even times of day may be relevant, depending on the topic.

Why?

In the Introduction of a scientific report, authors explain the reasons for conducting the research. With this background, it is usually clear why authors used the methods they did. However, occasionally the rationale for an aspect of the methods may not be readily apparent from the Introduction. In this situation, give the rationale in the Methods section. For complicated studies with many components, briefly explain the purpose of each component before describing it. Whether in the Introduction or in the Methods section, state clearly the assumptions underlying your use of a method. By making your assumptions explicit, you and your readers can assess their validity and more easily recognize the limitations of the research.

How?

Document the ways in which you implemented your methods. For example, routine yet critical aspects of studies should be described, such as mode of administration (for substances into bodies/organisms or surveys to respondents), consent procedures and other ethical matters, protocols for preventing sample contamination, incentives given to participants, and participant recruitment strategies, among other topics. Describe any type of special conditions necessary for or relevant to deploying your methods.

* * *

Because we report what we *did* in the Methods, we write the section in past tense. And because what we *did* is distinct from what we *found*, we do not include any results in the Methods section. One exception to this rule is that, in some fields, researchers describe the measurement characteristics of instruments in the Methods, generally for substantively-, rather than methodologically-, oriented studies.

Although it's important to report methods fully, do so concisely (see Chapter 29). If you used a method common to your field, simply mention it and give one or more citations to sources that describe it in detail.

Nonetheless, sometimes the Methods section is too long to be included in the main text of a scientific report. In this circumstance, create a "supplementary material" document to hold either a full description of the methods or just certain parts that we removed from the main text of the report. If the publisher of a report doesn't include supplementary material online, then put it in an independent, long-term archive, such as the Internet Archive or the Open Science Framework, and cite the web address for the supplementary material in the main scientific report.

Journals and fields vary on the precise details to be included in a Methods section and the format in which they should be presented. Those guidelines and my tips here can help make your research more understandable and reproducible.

20

Writing a Results Section

*We ... published without reserve all the results of our research ...
no detail was kept secret ....*
—Marie Curie, 1923

Along with the Methods section (see Chapter 19), the Results section is the heart of a scientific report. The Results are facts and descriptions of what you *found*. The only interpretations included in this section are objective—those which could be defined beforehand, such as how well an observed pattern corresponds to that predicted, perhaps in statistical, mathematical, or conventional terms.

By reporting your results completely and carefully, you permit readers to consider all of your evidence, and ensure that your work will be comparable to future research (see Chapter 14) and worthy of inclusion in systematic reviews (see Chapter 15). Report *all* of your results without fail. As you write this section, check your pre-registered protocol (see Chapter 4), analytic output (including images and graphics), lab notebooks, field notes, and other documentation to confirm you include everything. In fact, these materials are your ingredients for the text in the Results section. Just group the specific results in your collection by type and relationship with each other. Then, order the groups and the specific results within them, and draft one or more sentences to describe each result (see Chapter 18). Note explicitly in this section any results you exclude that would be expected from the protocol. Similarly, identify results as unplanned additions if you did not describe them in the protocol but include them in your report. Also check that you include every result expected from the Methods section. Amend the Methods section if it lacks a description of a method that underlies a result you report.

Report your results in detail. If you summarize a result (e.g., "we found no evidence for X") but don't give the details in the report, report them elsewhere, as in an appendix or online supplementary material (at the publisher's site or another stable location, such the Open Science Framework or the Internet Archive; see Chapter 19). Without such details, summaries of a result are essentially anecdotes, and readers may view them as attempts to deceive. Similarly, if an editor forces you to remove results in the form of text, tables, or figures from your report due to space limitations, put them in an appendix or supplementary material.

In any report that includes statistical analyses, always report univariate summary statistics on independent (predictor) and dependent (outcome/response) variables, including multiple measures of central tendency and dispersion for numeric data. Even if your focus is on multivariate relationships, also report bivariate (unadjusted) results, both to enhance readers' and your own understanding as well as make your work replicable (especially if your data aren't archived publicly). Multivariate results often are not replicable because other researchers do not study the same sets of variables. Always report *measures* of association, fit, or effect (see Chapter 8) and their associated uncertainties (e.g., standard errors or confidence intervals).

In the text of the Results section, avoid redundancy with the tables and figures. Summarize the main findings that appear in the tables and figures, perhaps highlighting one or a few of the embedded specific results. But leave out the full technical and statistical details, even for the highlighted specific results. Keep the text as words, without numbers or symbols, to the extent possible for brevity and clarity. If your sentences include many numbers and/or symbols, consider moving such technical and difficult-to-read content to tables or figures.

Write most parts of the Results section in the past tense, because the events and measurements in your study occurred in the past. However, there are exceptions. Once you transform your observations into data, the data persist (especially if you archive them publicly—see Chapter 9) and may even exist literally in your report. Sometimes you can describe patterns in or estimates from the data most naturally in the present tense, such as "these cases account for 58% of all observations", "the mean value is an overestimate by 10% because ...," and "in the figure, the coefficients tend to hover around 0." If you are uncertain which tense to use, just ask yourself whether the verb in your sentence refers to something that was clearly in the past. If you answer "yes," then use the past tense. Otherwise, it probably doesn't make much difference which tense you use, and the present tense may even be more accurate.

As much as you can, make the structure of the Results section parallel to that of the Methods section. Readers will have an idea of what you did from your Methods section, but their understanding can be lost easily if your presentation of the Results is at odds with the scheme you used for the Methods. Use one term only for each separate procedure, material, and measure throughout both the Methods and Results sections—no alternating between synonyms, which may be unfamiliar to some readers. Double-check! Many reports are uninterpretable because their authors did not organize and word these sections consistently.

Normally, the Results section does not include a description of the methods. Sometimes, however, authors must compromise, such as when a journal forces authors to put the Methods section at the end of the article (or, worse, in online supplementary material). In these cases, authors must summarize their methods at different points in the Results section to make their reports understandable. This format lengthens the report needlessly and undermines clear communication.

Some journals merge the Results and Discussion into a single section, at least for some reports. This makes it difficult for readers, and often authors, to know where facts end and interpretation and speculation begin. If you are forced to use a combined Results and Discussion section, focus on results exclusively in the first part of the section, and then discuss them in the second part. Make this division clear with a transitional phrase, such as "We interpret the results ..." or "I speculate that these results ...." Other journals may use sections with different labels. The key is to keep methods, results, and discussion distinct from each other, regardless of the headings. If you must use a format that does not allow *any* labeled sections, organize your material as if such headings were present, along with parallel transitional phrases that make clear the kind of content that follows.

21

Writing a Discussion Section

Bold ideas, unjustified anticipations, and speculative thought, are our only means for interpreting nature: our only organon, our only instrument, for grasping her. And we must hazard them to win our prize. Those among us who are unwilling to expose their ideas to the hazard of refutation do not take part in the scientific game.
—*Karl Popper, 1959*

For example, if you're doing an experiment, you should report everything that you think might make it invalid—not only what you think is right about it; other causes that could possibly explain your results; and things you thought of that you've eliminated by some other experiment, and how they worked—to make sure the other fellow can tell they have been eliminated. Details that could throw doubt on your interpretation must be given, if you know them. You must do the best you can—if you know anything at all wrong, or possibly wrong—to explain it ... the idea is to try to give all of the information to help others to judge the value of your contribution; not just the information that leads to judgment in one particular direction or another.
—*Richard Feynman, 1974*

In the Discussion section, you discuss your results. This means that you cannot write the Discussion faithfully until you know what your results are and have described them in the Results section (see Chapter 20). The Results section, prior publications, and your notes (see Chapter 46) give you much of the material you need for writing the Discussion.

There are several common elements in a Discussion section: summary of results, interpretation of results, generalization, limitations, and practical implications. While common, these elements are not universal across fields, studies, and types of research reports. Indeed, the only truly essential sections of a research report are the Methods and Results, although virtually all reports include at a least a token Introduction and Discussion.

Summary of results

The first paragraph(s) of the Discussion often is a summary of the main results. Use the same factual tone in this summary as you did in the Results section. For most reports, one paragraph is sufficient for the summary. If your report is very short, a summary may not be necessary, or could be just the first one or two sentences of a Discussion paragraph. In the summary and any other part of the Discussion, mention results only if you reported them in the Results section—do not include any new information about your study in the Discussion.

Interpretation of results

Describe all interpretations consistent with the results you discuss, including interpretations you dispute. Ideally, you will have already identified many of these interpretations when designing your research and preparing your pre-registered protocol (see Chapters 3 and 4). Consciously challenge yourself to find and mention alternate interpretations. One way to do this is to envision reactions to your results from likely critics.

When discussing an unexpected result, scientists sometimes blame their data, methods, or design as an excuse for the outcome, especially if it is inconsistent with their hypothesis or assumption. For instance, authors might explain the unexpected result as being due to error in the data, a problematic method, or inadequate design. However, methods and designs are known at the outset of a study. Why would researchers use methods or designs inappropriate for evaluating their hypotheses? Problems with the data sometimes cannot be anticipated. Yet these problems are a possible explanation only if they have been identified objectively in the actual data used in evaluating the hypothesis *and* are associated with the unexpected result. If problems with the data can cause unexpected results, they can also cause expected results. How often do scientists explain *expected* results as possibly being the product of problems in their data, methods, and designs? I don't think I have ever seen a case of such explanation. Blaming the data, methods, or design for unexpected results represents a double standard and illogical defense of an hypothesis. In fact, this fallacy renders the hypothesis unfalsifiable.

Interpreting results also involves putting them in context with prior research. Describe relevant previous studies briefly and note similarities and differences in results as well as methods and design. If your research involved the same research design and methods as the prior research, compare results formally and quantitatively, if possible. Otherwise, keep your comparisons brief and summarize prior work, relying on previous systematic reviews if available (see Chapter 15). The Discussion is not a narrative literature review. Listing who found what in tedious sentence after sentence does not enlighten. If prior research similar to yours showed different results, then give interpretations of your own results very briefly, if at all. It is pointless to interpret inconsistent results apart from speculating why they are inconsistent.

Assess relevant hypotheses and theories in light of your results, but with caveats about any weaknesses in your methods and design. Interpretations of results often include the seeds of new hypotheses to evaluate. State these hypotheses explicitly, and propose designs that can distinguish between competing explanations. Don't hide these as personal research plans—you will have more ideas than you can ever study, and the ideas you present may help others in their work.

In this and other parts of the Discussion, use verbs in active voice that make your uncertainty clear, such as "think," "believe," "interpret," and "speculate." Readers may mistake interpretation and opinion for fact when authors do not write accurately (see Chapter 27).

Generalization

In Discussion sections, and research broadly, scientists frequently make claims beyond the limits of generalization. Researchers studying a phenomenon that varies across space or time make this mistake when they generalize results from one time or place to another. Other common unjustified claims include the following:

study is based on:	but the claim extends to:
in vitro	in vivo
one species (e.g., mouse)	another species (e.g., human)
simulation	real world
one non-probability sample	whole population
one society	another society or all humans
lab/controlled conditions	field/nature/real world

These are differences in kind, not degree. The chasm between the domains on the left with those on the right cannot be bridged by inference or analogy.

Making claims within the boundaries of generalization is not a matter of modesty; it is a matter of rationality and accuracy. Moreover, a single study is a weak basis for generalizing in any case, even within the limits of generalization. A comprehensive systematic review (see Chapter 15) is the only good way to generalize within such limits.

Limitations

Almost every study suffers from limitations. Highlight constraints due to method or design (see Chapter 52) and possible errors and biases. In most research, regardless of field, measurement is indirect to some extent, and thus vulnerable to problems. Note specifically how each limitation might have affected particular results.

Practical implications

The practical implications may be the most read part of the Discussion, especially by scientists who are not specialists on the topic, journalists, and other non-scientists. Of course, some reports do not include practical implications, but even for basic (non-applied) research, authors often comment on the applied consequences of their work.

Although it is normative in many fields, avoid making optimistic or pessimistic extrapolations beyond the limits of generalization to matters of health, safety, environment, or other matters of popular concern. Many, if not most, such statements are self-serving: for example, promoting the authors' commercial interests, raising hysteria to maintain and boost funding for research in the field (especially by the authors!), or advancing authors' or funders' political agendas. Research reports with attention-grabbing claims, if publicized, can produce unreasonable fear or hope in the general public. We are obligated ethically not to mislead others about our research. Therefore, present the full range of *potential* implications, given the inevitable uncertainties in results, rather than just those in one direction. This is the equivalent of describing all interpretations consistent with a particular result, not just the interpretation you favor.

Be modest in recommending actions to take, whether of a practical or scientific nature. Scientists are not philosopher kings. Adjust the strength of your recommendations to the strength of your evidence and logic, using the verbs "should" and "must" sparingly, if at all. Single unreplicated studies, observational studies of cause and effect relationships, and extrapolation beyond the limits of generalization very rarely are a sound basis for a strong recommendation.

The practical implications are the logically and empirically weakest part of the Discussion and also the part most susceptible to our biases. Accordingly, use especially tentative language to reflect the varieties of extrapolation you have used in drawing implications.

Anything goes?

Although as author you have wide freedom in the comments you can make in the Discussion, exercise your liberty cautiously. In the Discussion, avoid promoting your work and topic as highly important, congratulating yourself repeatedly for doing something first or better than before, or signaling your ideological virtues. Such acts have nothing to do with science. Also, speculating in the Discussion is not a license to engage in logical fallacies and unsound reasoning (see Chapter 56). Write the Discussion with the same emphasis on scientific integrity that you used in the Methods and Results.

22

Writing an Introduction

Introductions are supposed to introduce. How can you introduce something you haven't written yet? You don't know what it is. Get it written and then you can introduce it.
—Everett Hughes, c. 1950

The Introduction is easy to write if you draw on the rest of your already drafted report, along with your pre-registered protocol (see Chapter 4), prior work you identified in your literature search (see Chapter 5), and notes you made during your research (see Chapters 18 and 46).

The only essential elements of the Introduction are presenting the topic and identifying the problem(s) you addressed in your research. Other elements that commonly appear in Introductions are not constant across fields or authors, and are matters of tradition and taste. Nonetheless, including at least some of these other elements helps readers understand *why* you did the research.

Jeffrey McDonnell recommended that authors answer three questions in the Introduction: "What is the status quo? What is wrong with the status quo? How does this new paper go beyond the status quo?" Use these questions to organize most of the Introduction.

In describing status quo, don't recapitulate the history of your field or topic. While this might be appropriate for a dissertation, it is unnecessary for and detrimental to a research report. Furthermore, the Introduction is not a literature review nor must it include one. Summarize the status quo briefly on the basis of sound systematic reviews (see Chapter 15), if available. Otherwise, note the range of prior findings, citing representative examples (see Chapter 26) and explaining that a formal summary awaits a systematic review. Avoid pseudo-summary statements like "most researchers found X but a few found Y," as they are usually meaningless and can be deceiving. The results might not be as mixed or as lopsided as they appear once the literature has been searched thoroughly and the methodological and statistical details have been considered. If there is no prior research on your topic or no previous studies like yours, note it explicitly, so readers know that you are not ignoring relevant work intentionally.

In describing what is wrong with the status quo, focus on the *main* problems, as you see them, dispassionately and objectively. Similarly, in describing how your work goes beyond the status quo, don't overstate the originality of your research or highlight trivial differences from prior work. Imagine how others who have studied your topic would react to your descriptions— would they agree? It is better to present your work accurately as a modest contribution than claim unjustified significance.

If you do not imply the general structure of your research elsewhere in the Introduction, then outline the specific parts and their rationales in the last paragraph of the Introduction. This is useful when there are unusual methods or multiple stages or studies in the research you report. The goal here is to ensure the reader knows why you did each major part of your research. You have the option instead to describe the rationales for these parts in the Methods (see Chapter 19). In any event, do not trace the conventional structure of a research report in the Introduction, such as "next, we describe our methods, then present our results, and in the final section discuss our findings." Readers already expect you will do this.

Also include the assumptions underlying your research in the Introduction (or Methods, as you see fit). By making your assumptions explicit, you strengthen the rationale for your design and methods and communicate the logic of your work more clearly.

Many authors include excessive material in the Introduction. Foreshadowing results and conclusions needlessly lengthens the report—you will summarize them in the abstract (see Chapter

23). Similarly, don't make points that stem from your results. Instead, write the Introduction as if you did not know your results. Also, avoid justifying your topic or methods with appeals to popularity (e.g., "increasingly studied," "widely used"). You don't need an excuse to study anything; justify your methods on rational grounds. Finally, keep the Introduction, and the rest of your report, free of emotional appeals, political statements, self-promotion, and hyperbole.

23

Writing an Abstract

Whatever is worth saying, can be stated in fifty words or less.
—Stanislaw Ulam, 1909–1984

An abstract is simply a summary of your report. Many more people will read the abstract than will ever read the main text. Therefore, spend extra effort to make it is as accurate and effective as possible.

Write the abstract after drafting the whole report. An abstract is not a plan for a report; writing it first would constrain the content of, and your orientation to, the report.

Begin by applying Ulam's dictum. Write a brief one or two sentence summary of your report. Imagine another scientist asks you to summarize your research. A simple answer would include what you did and what you found, and perhaps indicate why it is relevant. We scientists give brief summaries like this without difficulty when we talk informally with each other. This brief summary doesn't need to be included in your abstract, but writing it *before* you draft your abstract is very helpful. It focuses your mind on the essence of your contribution and prevents your veering into irrelevant matters.

Even if there is no specific structure required for the abstract of your report, follow an implicit structure anyway. The structure guides you to include the essential information and gives a natural flow to the abstract. The basic elements, in order, are: background/objective/rationale; methods/material/design; results; and interpretations/conclusions/implications. You don't need to label these sections (unless it is an explicitly structured abstract). In my opinion, the last section is optional, as clearly stated results often are a sufficient conclusion. Conclusions that consist of results restated in a slightly different way are a mindless waste of authors' and readers' time.

The content of the abstract should not deviate from or distort the content in the report in any way. The abstract should be derived completely from the main text of your report. You have already summarized aspects of your research at different points in your report—just reuse sentences, or parts of them, from the main text that you have already written. There is no sense in trying to summarize your work anew. Usually you can simply copy a few topic or summary sentences from the sections in the report corresponding to the sections in abstract. The first paragraph of the Discussion (see Chapter 21) often has good sentences for most or all sections of the abstract. Through this process you can assemble a small amount of raw material—ideally less than a page of text—from which you can compose the abstract. It may be necessary to condense and reword these sentences, but by limiting the scope to these sentences, you make the task much easier.

When writing the abstract, summarize your primary results according to your pre-registered research plan (see Chapter 4), no matter what they are and even if they are counter to your expectations or preferences. If you include secondary results in the abstract, mention them only after describing the primary results. Also, avoid empty statements that refer to expected material, such as "We conducted an experiment to test this hypothesis," "We describe the results from this study," "We discuss the implications of these results," "More research is needed on X," and "We offer suggestions for future research."

24

Writing Acknowledgments

No man is an island, entire of itself.
—John Donne, 1623

... praise the bridge that carried you over.
—George Colman the Younger, 1797

The authors of a scientific report typically are not the only persons who did the research, wrote the report, and made the work possible. In the acknowledgments section of a scientific report, you give a full accounting of the others involved and their contributions. With this information, readers can know how you were able to accomplish what you did, which can also allay their concerns about potential fraud or theft of others' work. Give credit to those who deserve it. Don't take credit for tasks you did not do or ideas that others gave you, thereby preventing misrepresentation. Thank those who helped you, which is a basic courtesy, but also is essential for developing and strengthening relationships. Include everyone who assisted with different aspects of the work, such as data collectors, programmers, technicians, editors, writers, research partners (see Chapter 37), and colleagues who reviewed the report, among others, even if you paid them.

Unless you note it elsewhere, it is crucial to report the funding source specifically in the acknowledgments. Even if you received no funding for the research, state that fact explicitly. It is difficult for readers to know whether there was funding if you do not mention the topic.

As you do your research, keep notes on who has helped you and how. This enables you to remember everyone whom you should acknowledge. Recognize individuals by name whenever possible rather than mention groups (e.g., "members of the

Smith laboratory" or "participants in the weekly seminar series") or individuals by their roles (e.g., "the laboratory manager" or "principal of the school"). Always take special care not to omit low status persons and non-scientists. Conversely, don't mention high status persons whom you otherwise wouldn't to curry favor or suggest powerful connections. Just like other sections of a scientific report, the acknowledgments should be about facts, not politics.

25

Figuring

If statistical graphics, although born just yesterday, extends its reach and circle of applications every day ... it is because it replaces long tables of numbers and it allows one not only to embrace at a glance the series of phenomena, but also to signal the correspondences or anomalies, to find the causes, to identify the laws.
—Emile Cheysson, 1880

Graphs, pictures, videos, and other images are integral parts of scientific communication. Figures allow you to convey information that would be difficult or impossible with words alone. In many situations, figures also let you communicate with less space and more clarity than text or tables. Figures, of course, are also useful during study design and data analysis even if you choose not to include them in your reports and presentations. Visual displays permit you to represent hypotheses and study procedures effectively, and see patterns in your data that you might otherwise miss.

Many guides on designing graphs and other figures cover the fine details of visualization in scientific research. Instead, I focus here on very basic, common, and practical aspects of figures that researchers often overlook.

Message

Perhaps most important, have a point that you communicate with your figure. Scientists sometimes present pretty figures that look interesting but convey no relevant scientific message. An irrelevant or bad figure is worse than no figure. Every report does not require a figure (or a table). Include just those figures and tables that are essential and not redundant with each other or the text.

A picture may be worth a thousand words, but if a figure requires a thousand words (or even two hundred) to explain, it is probably not a good choice. Use figures that can be understood easily with only a little explanation.

Simplicity

As with other aspects of research (see Chapters 13 and 29), make your figures simple in all aspects. Remove all unnecessary lines, numbers, text, boxes, shadings, patterns, and effects. This often requires a special pruning effort, as most graphics software adds unnecessary and distracting features by default. Design your figures so that the viewer can grasp immediately—within a few seconds—the point(s) you seek to communicate. Some figures may require extensive study yet still communicate important points. Put such graphs in an appendix or supplementary online material, as they interrupt the reader's train of thought, hindering memory of preceding material in your report.

Researchers sometimes produce complex figures to represent hypotheses, theories, models, and processes with many boxes and arrows. Such figures often reflect sloppy thinking. Complicated conceptual frameworks are little better than raw data, so be especially self-critical when making figures of them.

Simplicity includes economy. Use as few figures as necessary. Sometimes authors embed many figures within a single figure (such as Figure 1a, b, c, d, e, f, g, and h) that has a very long caption. Multi-part figures may be attempts to bypass journal limits on the number of figures in a report. Nonetheless, each part of such a figure tends to be tiny, masking relevant details. Moreover, long captions sap readers' attention as they try to find a particular sub-caption and its corresponding graphic and then alternate between the two as necessary before repeating the cycle for another part of the figure. Often readers must return to reading the main text before viewing another part of the figure, further taxing their attention and memory. Therefore, avoid multi-part figures and publication outlets that effectively force authors to use them.

Human perception

Create figures that match human perceptual abilities. All essential components of a figure should be large and legible, with high contrast. Graphical display strategies vary dramatically in their effectiveness to convey information to human viewers accurately. In descending order of effectiveness, strategies for displaying ordinal, interval, and ratio scale data are as follows:

- position on a scale (e.g., scatter plot, box plot)   *most effective*
- length in one dimension (e.g., simple bar chart)
- angle or slope (e.g., pie chart)
- area (e.g., bubbles in bubble chart)
- depth in three dimensions (e.g., three-dimensional graphs)
- shading intensity (e.g., gray scale)
- curvature (e.g., comparing curves in the same plot) and volume (e.g., three-dimensional graphs)   *least effective*

Adding three-dimensional effects to one- or two-dimensional graphs, such as to an otherwise simple bar chart, degrades effectiveness substantially.

Strategies for displaying different values of categorical or nominal data, in descending order of effectiveness, are as follows:

- position in space (e.g., data points or   *most effective* summaries in different sections of a graph)
- highly distinct color hue (e.g., blue, yellow, red) and shade (e.g., white, gray, black)
- shape (e.g., triangle, circle, square;   *least effective* solid, dashed, and dotted lines)

Color hue, shade, and shape are limited, in practical terms, to data with no more than 4 categories. Otherwise, viewers cannot remember their corresponding symbols and distinguish between them easily. If viewers need to refer to a legend or key for the symbols repeatedly to understand the figure, many will give up and not receive the message of the figure.

Color hue is a bad strategy for representing ratio, interval, or ordinal scale data. A heat map or color map with a rainbow continuum of color hues is an example of this approach. The rainbow continuum, although familiar to many viewers, is not intuitive to most as a quantitative scale, as even the sequence of colors requires effort to re-learn and remember. Human viewers are also especially vulnerable to perceptual illusions caused by contrast effects from the hues and shades of adjacent data points, distorting interpretation. Color hue actually is a poor strategy for representing any type of data. Color blindness affects up to 10% of males and smaller proportions of females in many populations. Color images in publications also may ultimately be rendered in gray scale when readers photocopy or print them.

Denotation

Very few figures are truly self-explanatory. Explain the main content of the figure in the text, figure title, and/or caption. Label axes, show scales, specify what error bars signify (e.g., confidence intervals, standard errors, standard deviations, etc.), and indicate what different shades, shapes, lines, contours, super- and sub-scripts, color hues, and other symbols represent (such as in a legend or figure footnotes). Note whenever you reuse or modify an image, in whole or in part, from another source, even if it is your own, and cite the original.

Integrity

Visualization and all other aspects of scientific communication are about *describing*, not *persuading*. Display results honestly. Avoid the many tricks some data analysts use to deceive. For example, sometimes researchers magnify patterns by removing much of the interval on one axis, showing the extreme end only.

Although this can be done validly by showing a symbol on the axis that you removed the interval, the resulting image still leaves the viewer with a distorted view. Instead, simply show the whole interval for every axis. Another deceptive practice is to restrict the range of the graphed data. By displaying a subset of the data only, researchers can create impressions of associations (including rises and declines in temporal data) or lack thereof when the full range of data would show otherwise. A cardinal rule of data analysis, including visualization, is to analyze all of the data (see Chapter 8). Thus, show all of the data in your figures. You can focus on subsets of the data once you have given the full context. A further common example of a misleading graph is to show, over time, cumulative totals of some phenomenon (e.g., births, deaths, emissions, resources consumed, sales, etc.) which must grow or at least not fall. Such a graph implies increased rates or frequencies even when the trend is actually decreasing. A better approach is to plot the quantity for each time point and include the cumulative graph as a secondary figure, if at all.

26

Citing Sources

I shall never be ashamed to quote a bad author if what he says is good.
—*Lucius Annaeus Seneca, 4 B.C.–A.D. 65*

There is not less wit nor invention in applying rightly a thought one finds in a book, than in being the first author of that thought.
—*Pierre Bayle, 1647–1706*

In scientific and other formal writing, it is crucial for authors to cite sources for ideas and facts that they did not discover or observe themselves. Quite simply, citing sources correctly is about giving credit where credit is due.

If we, as authors, do not cite our sources, why should anyone else? If we don't search the literature well (see Chapter 5) and cite our sources, why should other authors pay attention to our work and cite it?

Apart from this fundamental golden rule of citation, there are other reasons why citing correctly helps authors. First, authors can embarrass themselves by claiming something as new when it isn't, or by ignoring clearly relevant ideas or findings.

Second, authors' literature searches and citation practices reflect on their scholarship. If an author neglects pertinent prior work or cites in a biased way, readers may well suspect that the author is lazy, careless, or unfair. Readers might think the research the author reports is similarly flawed.

Third, each scientific report, whether an article, book, or other publication, is a stepping stone in the progress of science. If authors don't link their scientific reports to the past properly, they become stones out of reach from the others and do not help readers. By making connections to relevant past work, authors also demonstrate their knowledge of the science underlying their work.

Follow six essential principles of good citation:

Cite original sources

Strive to cite the first work describing a particular idea or result. The original conceiver or discoverer naturally deserves the most credit. Keep in mind that often, the author(s) popularly perceived as the first is, in fact, neither the first nor the most deserving. Correct the historical record, if possible, no matter how many generations of scientists have misidentified or ignored the original source.

Appreciate history

Some authors and editors artificially limit the scope of their literature searches and citations to very recent work, such as the previous 5 or 10 years. Even in fast developing fields, there often is older work that is pertinent and deserves mention. Consider citing relevant work no matter when it was published. More recent research is not necessarily better than older research, and citations only to recent work may sometimes indicate an author's short-sightedness or laziness.

Cite representatively

It is easy to skew a reader's impression of past research by citing some and not other work. Too often authors cite selectively — cherry picking their sources — perhaps to promote, consciously or unconsciously, a particular view. Some authors find it hard to cite their rivals, researchers whom they do not like, or work with different results than their own.

Despite these impulses, it is possible to develop more balanced citation habits. Treat citing sources as any other scientific task, such as making observations: dispassionately, focused solely on the facts and not emotions. Take it as a challenge to cite rivals, those who have done similar prior work, those with whom you disagree, or those whom you dislike. If prior results or opinion are mixed, cite all sides. When writing and citing, it can be helpful to imagine whether a critic might consider the sources we cite as representative, and adjust accordingly.

Cite accurately

Several of my colleagues and I have found that many, if not most, citations our work receives are inaccurate. That is, other authors credit us with findings and ideas that we did not describe, and surprisingly often cite our work to support points that our work directly contradicts. Such citations are cancers on the web of scientific knowledge and reflect poorly on the citing authors. It is better not to cite at all than cite incorrectly. Incorrect citations probably stem most often from authors not actually reading the original sources they cite, a practice which unfortunately may be widespread.

Online sources tend to disappear over time, due to the especially impermanent nature of the World Wide Web. Use a web archiving service, such as archive.today, to preserve online sources and links to them for the future.

Cite according to content, not prestige

Ideally, our decisions about which works to cite should rest on their content alone. The prestige of the author or journal, or the type of publication (such as journal article, book, book chapter, thesis, dissertation, unpublished report, etc.) are irrelevant. Some authors use citations as political tools, to curry favor with reviewers and editors, or align themselves with one group of researchers or another. Such tactics are more prone to biased citation than an approach focused on ideas and evidence.

Cite reviews

Typically, we have limited space in which to discuss and cite previous work. When describing prior empirical research, it is best to cite comprehensive meta-analyses or systematic reviews, or if these are not available, narrative reviews, instead of individual reports. (Although when you do cite individual reports, follow the other principles in this chapter.) Not only do such reviews reduce the number of works cited (which is important if we cannot exceed a certain number of references in a report), but reviews provide the most reliable summaries of the corresponding research if done well (see Chapter 15).

27

Using Accurate Language

The difference between the almost right word and the right word is really a large matter — 'tis the difference between the lightning-bug and the lightning.
—Mark Twain, 1888

We scientists may claim to know that correlation does not imply causation, but our language often indicates otherwise. Causation can be inferred only from properly controlled experiments (see Chapter 52). Despite this, researchers frequently use causal terms (such as "cause," "effect," "impact," "modulate," "induce," "influence," "determine," etc.) to describe associations in observational research. Still other terms (such as "reduce," "enlarge," "decrease," "increase," "enhance," and "promote," among many others) also imply causation when used as active verbs for describing the association between variables (e.g., "X reduced Y"). Instead of using causal terms inappropriately, use terms that accurately reflect the observational nature of the data, such as "correlation," "association," and "relationship" (or their verb equivalents). "Risk factor" and "protective factor" are just unnecessarily fancy words for "correlate" (used as a noun). These terms are often perceived to imply a potential or likely causal relationship, and consequently they are problematic words. There are, however, some terms that imply causation that may be difficult to avoid using in reporting observational research. For instance, some statistical terms involve causal language (such as "effect size") even though they can be applied validly to analyses of non-experimental data.

It is easier to think about causal relationships than mere correlations. We humans are by nature inclined to see cause and effect where none exists. Thus, it can be helpful to conceive of relationships as causal when making hypotheses and developing tentative interpretations of results. Just ensure that in describing such hypotheses and interpretations you note that the causation is speculative, not demonstrated.

Other common types of inaccurate language stem from scientists taking shortcuts that, intentionally or not, misrepresent research. When discussing empirical results, refer to populations (of people, other organisms, or other objects of study) only when they have been sampled with a method (usually probability-based) that ensures the sample is representative of the population as a whole. Authors often lapse into describing non-representative samples as populations, without the essential qualifiers about the nature of the sample. The type of measurement also is a key detail for communicating results meaningfully. For example, when discussing results about people, distinguish between subjective measures of phenomena, obtained through self-report, from objective measures of phenomena. Similarly, make clear when results are based on proxy measures (such as a test associated with, but not strictly diagnostic of, a disease) rather than the fundamental variable of interest (such as the disease itself). If we confuse concepts with their measures, we might make claims that extend beyond the evidence. Furthermore, scientists in many fields have the regrettable tendency to use the term "bias" when they mean "error"—so much so, that several basic methodological concepts in these fields have been misnamed. Use "bias" only when referring to systematic error—that is, error that tends toward one direction.

Some writing styles also result in inaccuracy. Strive to use the active voice ("we collected data"), rather than the passive voice ("data were collected"), not just in methods sections (see Chapter 19), but in scientific writing generally. Although there are some circumstances when the passive voice is better, usually the active voice is more accurate and honest. The active voice indicates the interplay between real subjects and objects (in the grammatical sense), while in the passive voice events often seem to occur as if by magic. Researchers often use the passive voice to make their actions invisible to the reader, which can give the air of objectivity. However, objective descriptions of research by definition include a transparent and full accounting of researchers performing those actions. The passive voice obscures who did what and how events actually happened.

Authors sometimes mislead by using the pronoun "we" instead of "I" in single authored papers. When this happens, it is not clear whether there are uncredited co-authors, or the author is expressing himself or herself with the "royal we" as some sovereigns or religious leaders do. Some authors use the "author's we" instead of "you" or "one," as in "surveying the literature, we find …." It is always possible to rewrite such sentences without "you," "one," or "we". Use "we" when there are multiple authors, but otherwise avoid this pronoun in scientific writing.

In recent decades, researchers have succumbed to other forms of inaccuracy. For instance, many verbs apply to living subjects only. Yet sometimes scientists mistakenly state that studies, models, and methods, for example, "try," "ask," "seek," "assume," "infer," "look," "find," "argue," or "interpret." Only people can do such things in the context of research. In your writing, make sure that human actions are tied to human subjects (in the grammatical sense), and not inanimate subjects.

Dictionaries and thesauruses are essential tools for writing accurately, no matter how large your vocabulary. Consult these references whenever you have any uncertainty about the precise meaning, connotation, or tone of a word. I turn to a dictionary or thesaurus once or more when writing a typical paragraph.

We scientists attempt to be accurate in doing research. But if the language we use in reporting our research is inaccurate, we hinder clear communication and subvert the accuracy in our research.

28

Writing across Generations

The men of old changed the name of their methods from problem to problem, so that as no specific explanation was given, there is no way of telling their theoretical origin or basis.
—*Yang Hui, 1275*

See, I cast the die, and I write the book. Whether it is to be read by the people of the present or the future makes no difference: let it await its reader for a hundred years, if God Himself has stood ready for six thousand years for one to study him.
—*Johannes Kepler, 1619*

[Books] are the voices of the distant and the dead.
—*William Ellery Channing, 1838*

Good scientific writing is classic, in most senses of the word: simple, elegant, memorable, high quality, standard, and timeless. The last two senses are particularly critical. When different scientists use the same words to refer to the same things, the chances for good communication increase. And because scientists attempt to communicate across generations and cultures, both among the living and the not-yet living, standardization of terms must be constant and timeless. Some religions have been successfully propagated across millennia and cultures, in part, because their core writings have remained fixed in classical versions of languages, such as Hebrew, Latin, and Arabic. Science requires the same fidelity to maintain the accessibility of knowledge. Indeed, scientists used classical languages, especially Latin, to communicate their research for many centuries.

A few simple practices can help make your scientific writing understood by readers in the future. Write as if you are writing for scientists of your grandparents' generation. If past generations could understand your writing, it is more likely that future generations will also. Define non-standard words and avoid slang. Neologisms are necessary for truly new concepts only. Coin new terms (or extend new meanings to existing terms) only after searching the literature thoroughly and consulting with colleagues to check that there are no other suitable ones already in use. Hold fast against politically- and ideologically-motivated renaming of concepts, especially euphemisms, and ignore attempts by researchers to rename established concepts under the guise of making (false) intellectual contributions. Use terms from the language in which they were first expressed, or mention them alongside translations. Use conventional spellings, punctuation, and grammar—follow old and established styles rather than new ones. When multiple terms are commonly used for the same concept, list the synonyms as a service to the reader.

Reading the scientific literature from generations past (see Chapter 47) can give you a good sense for what is required to make your own writing intelligible to future readers. Notice how previous authors did or did not make their messages clear to you as a reader, and adjust your own writing accordingly to imitate what they did right and avoid what they did wrong.

Novelty and fashion are bad for long-term communication. When it comes to scientific writing, be a traditionalist.

29

Less is More

A multitude of words is no proof of a prudent mind.
— Thales, c. 624 B.C.–c. 546 B.C.

My liege, and madam, to expostulate
What majesty should be, what duty is,
What day is day, night night, and time is time,
Were nothing but to waste night, day, and time;
Therefore, since brevity's the soul of wit,
And tediousness the limbs and outward flourishes,
I will be brief.
— William Shakespeare, Hamlet, c. 1599

… it is excellent discipline for an author to … say all he has to say
in the fewest possible words, or his reader is sure to skip them; and
in the plainest possible words or his reader will certainly
misunderstand them.
— John Ruskin, 1857

In scientific writing, short is sweet. Brevity benefits both readers
and authors. For readers, concise writing reduces the time and
effort they must spend reading. For authors, writing concisely
helps to increase the clarity and logic of their reports, and allows
them to meet word limits for manuscripts, grant proposals, and
abstracts with ease. Furthermore, verbose writing is a public,
enduring sign of an author's sloppy work and fuzzy thought.

Word length

The goal in good scientific writing is to communicate well, not
show off with fancy words. Common and short words therefore
tend to be better for communication than uncommon and long
words with the same meanings. For example, as a verb, "use" is
better than "utilize" or "employ", and as a noun, "use" is much

better than "utilization" or "employment." Technical and medical jargons often include long terms that have short, non-technical equivalents. Whenever possible, use the non-technical word, perhaps after noting the jargonistic counterpart in parentheses.

Sentence length

Long sentences usually reflect confused or convoluted thoughts. Any sentence longer than 35 words automatically deserves scrutiny. In most cases, such sentences can be shortened or split in two. Short sentences also emphasize points.

There are many ways to shorten sentences. Cut empty phrases that carry no substantive meaning. Examples include "In order ... ", "It is important to note that ...", "I should mention ...", "We hasten to add ...", "Many studies demonstrate ...", "To be clear ...", "It hardly needs to be stated ...", and "Research has shown that ...". The text that would follow each of these can stand on its own. You don't need to say you are about to say something—just say it.

Put the subject and its verb as early as possible in a sentence. Putting both within the first 12 words makes it much easier for the reader to track the structure and meaning of a sentence. It also makes it easier, as you write, to see the natural break points in sentences that have become too long.

Use verbs rather than nouns to express actions. For instance, write "We analyzed X", instead of "We carried out an analysis of X"; write "I hypothesize that ...", instead of "I offer an hypothesis that ...." Also, use active voice as it tends to be shorter and more accurate (see Chapter 27) than passive voice.

As authors, sometimes we fall in love with particular words or phrases and then try to build sentences around them. This is the wrong direction for scientific writing. Decide on the point of a sentence and *then* pick the words that express that point best. If your beloved word or phrase isn't among them, so be it.

Avoid putting statistical results in the text, even within parentheses, when the same results are in a table or figure. It is fine to highlight a *very small* set of particular values in the text, but often this is not necessary. Similarly, put other technical material, such as chemical formulations/reactions, primers, mathematical formulas, detailed specifications, and so on, in tables, figures, or appendices whenever possible. These details often make sentences long and difficult to read.

Technical terms and other jargon that have no common language equivalents can reduce the number of words needed to describe something. Jargon is useful only if your audience is likely to know the terms and you define them at first mention.

While brevity is good, make sure not to sacrifice clarity in achieving it. For instance, scientists are prone to writing noun and adjective stacks—phrases containing many nouns or adjectives in succession—such as "field data collection procedures". Adding prepositions or converting nouns to verbs often makes these phrases easier to understand (e.g., "procedures for collecting data in the field"), which is worth the price of a few extra words. Of course, some noun stacks are conventional terms (e.g., "product-moment correlation coefficient") and it is best to use such terms given their standard and wide use.

Many authors also try to decrease sentence length by using acronyms frequently. Acronyms tax the reader's memory heavily and hinder understanding. I recommend using no more than three acronyms in the text of a report, no matter how long. Exceptions include common acronyms outside of scientific research; acronyms that function as proper nouns, such as names of projects or groups of researchers; and acronyms that are abbreviations of names, such as of organisms, viruses, molecules, minerals, equipment, and materials. Introduce an acronym only if you will use it several times in your text. It is pointless to use an acronym only once. Acronyms used a few times only multiply the cognitive burden for readers, who must spend extra effort and time to learn and remember the association.

Moreover, fewer words are better only when they involve fewer total syllables than another wording. Otherwise, the cognitive effort and time required to read and understand the text may actually increase.

As you polish your text, read each sentence slowly. Consider rewordings that would make it shorter. Also consider whether each word is necessary. Use Howard Becker's method of taking out words to see whether the meaning changes (or changes appreciably). If it doesn't, then leave the words out.

Paragraph length

Long paragraphs are exhausting for readers. Make each sentence count and hold your readers' attention. If a paragraph is longer than one page (letter/A4 size, 12 point font, double spaced, 1 inch margins all around), it is too long and should be shortened or split. Even paragraphs shorter than a page can be too long, depending on the context.

Keep paragraphs lean with the following tactics. Look for and eliminate repetitions of points. Make a point clearly once — that's enough. Similarly, after giving a detailed description of a subject or object once, it is not necessary to repeat it in the sentences and paragraphs immediately following. For instance, the "method for measuring X" can thereafter be "method", as long as you discuss no other methods in the same context. Also, avoid empty generic statements such as "results will be discussed" or "more research is needed" that indicate normative structures or expectations.

Avoid footnotes. They interrupt the flow of your presentation and distract the reader's focus. With footnotes, readers must decide whether to jump around in your report to get the extra information. If you want to give additional information, include it in the main text or put it in an appendix. Tangential comments can usually be cut entirely. Respect your readers by limiting interruptions and jumps.

Sometimes it is not possible to integrate a particular sentence or two into another paragraph without making it awkward. In such cases, create a new paragraph with the sentence(s). Ignore style rules about the minimum number of sentences in a paragraph, as they do not necessarily allow for clear communication. A paragraph is a set of sentences related to the same idea or ideas. If a sentence doesn't relate to the ideas in other paragraphs, it belongs in its own paragraph. Of course, if you have many single sentence paragraphs, it might be better to group them together into multi-sentence paragraphs under more general themes.

Manuscript length

There still may be excessive material in a manuscript even if the lengths of words, sentences, and paragraphs tend to be short. Pare your presentation to the fewest possible ideas and lump similar ideas together.

No matter how important a point you want to make is or how much time you spent developing it, if it doesn't fit well and isn't essential, cut it. Be merciless. In my own writing, I know I am beginning to show discipline when I feel uneasy cutting points I want to make that are nonetheless unnecessary. I keep all the text I cut in a separate document (see Chapter 18). This way, the points are not lost. Once the points are gone from the main draft, it is easy to forget about them and realize they weren't necessary.

In general, repetition is not good. A research report is a record of work and thought, not a persuasion piece. Avoid restating or rehashing content from an earlier section of the manuscript. If absolutely necessary, restate in a few words the topic and where you discussed it, and assume the reader has read or will read the other parts of the report. It is impossible to write for a reader who will read an unknown fraction of a report, or who cannot remember or consult earlier sections.

Long reports and books sometimes are necessary. But even in such works, the "less is more" principle applies. Break material into short sections (ideally, fewer than 1,000 words), and use headings and sub-headings to mark them. Relevant tables and figures can help prevent long blocks of text and give cognitive and visual relief to readers.

General strategies for developing brevity

The present letter is a very long one, simply because I had no leisure to make it shorter.
—*Blaise Pascal, 1656*

Writing less is more work. With practice, however, it gets easier to write concisely. There are several general ways to make writing concisely a habit.

Write letters to the editor of scientific journals. Such letters typically have severe word and reference limits, which force authors to distill their arguments and trim unneeded words. Start by drafting a letter that includes all of the points you want to make. Hopefully the word count of your draft exceeds the journal's limit, so you can learn from the exercise. Then use the tactics for shortening that I have described in this chapter.

Work with other researchers who write concisely. As you write papers together, observe how they write and edit, and take their writing criticisms to heart. This doesn't mean you must follow all of their suggestions, but their feedback will help you internalize the process of evaluating the necessity of every word and point you write.

Similarly, study the writing of those who write well and concisely. And if you must read the writing of a verbose author, just skim it to glean the essential content.

Read books for children who are just learning to read (I recommend those by Dr. Seuss). Authors of such books often have mastered the simple style that is ideal for good writing of any type, with short words, short sentences, short paragraphs, short overall length, clarity, and logical structure. I relearned how to write, in part, while reading with my kids—the same books, over and over—when they were young.

Write about scientific topics for non-scientific audiences. Such writing presses you to write concisely, make your assumptions clear, and abandon jargon and complex descriptions of technical details.

Number of publications

Because the academic career puts a young person in a sort of compulsory situation to produce scientific papers in impressive quantity, a temptation to superficiality arises that only strong characters are able to resist.
—Albert Einstein, 1955

All scientists face finite limits of time, energy, and mind. We can divide those resources into few high quality publications, or many low quality publications. The impulse to publish frequently often leads to redundant, trivial, incomplete, and/or contradictory reports. Publishing frequently also tempts researchers to cut corners on good research practices and good writing.

We scientists are obligated to report our research coherently and completely (see Chapter 32). Report a whole study in a single publication whenever possible. This might mean publishing a monograph or book rather than a series of articles. Reporting different, often overlapping, aspects of a study in separate articles fragments the literature. Readers may wonder whether the multiple reports are from the same or different studies, if they even can find and read all of the reports. Separate reports leave inconsistencies between them unexplained.

Some researchers also write the same paper repeatedly, covering very similar points from one to the next, whether with new results or old. Such repetitive works most often occur in narrative reviews, commentaries, so-called think pieces, and other unstructured publications, and typically don't include citations to their clones. Some authors regard their duplicative publications as attempts to reach different audiences—in effect, as marketing efforts. Yet the scientific literature is open for searching and reading by all researchers (see Chapter 5), making repetitive publication unnecessary. Filling the literature with advertising pollutes it, wastes readers' time, and reflects poorly on the authors who produce the redundant publications.

30

Checking—Part II

... you will find it a very good practice, always to verify your references, sir!
— Martin Joseph Routh, 1847

Proofread carefully to see if you any words out.
—William Safire, 1980

A crucial near-final step in preparing a research report is to check it for completeness, grammar, punctuation, and accuracy. In some fields, there are common checklists for reporting different kinds of studies. If a reporting checklist applies to your research, use it as a guide. Not every item on such checklists necessarily applies to every study in their corresponding categories. Still, they are useful to ensure you have included important information in your report or have a good reason for not doing so.

Always proofread your reports. The automatic spell check in a word processor is *not* a replacement for proofreading, as not all correctly spelled words may be the ones you intended or make sense in their contexts. Moreover, many scientific and technical terms are not in word processor dictionaries. When proofreading, make sure you have defined all acronyms and abbreviations (unless the latter are standard) when you first mention them in your report. Don't rely on journal reviewers and editors (even the journal's copyeditors) to proofread and edit your manuscripts. Submitting poorly prepared manuscripts could harm your reputation and lower your chances of publication. If the journal offers limited or no copyediting, it is possible a mistake-riddled report could be published. When you publish your work in preprint archives (see Chapter 32), *you* are the copyeditor, which makes proofreading especially important to do well.

Proofreading is not a one-time act. As you draft a sentence, you may proofread it many times, each time mulling the wording and intent. Then after composing a paragraph, you may proofread it many times more, considering how to modify it each time. After drafting a whole manuscript, proofread it several times, with the first passes focusing on broad issues and later passes focusing on finer points and details. Before seeking comments on it from others (see Chapter 31), I recommend taking a break from the manuscript—by engaging in a non-scientific activity for many hours, sleeping over night, or waiting several days—before proofreading a final time. The break refreshes your view and wipes out ruts of attention and memory. This allows you to see problems you didn't before and improve your writing further.

If you don't use a reference manager (software) for your citations and references, then always check that every source cited in the text is listed in the references, and vice versa. Also, if you archive your data (see Chapter 9) or software publicly, confirm that you obtain results with the archived versions identical to those you report.

By checking in all of these ways, you not only ensure that your report is complete, grammatical, and accurate, but also demonstrate the care you take in your work. Researchers who do not check their reports may raise readers' doubts about the quality of the underlying research. If a scientist isn't careful about those aspects of their work that are public, how can others be confident that the scientist handled appropriately those aspects of research that are not public?

31

Soliciting Comments

Take the advice of a faithful friend, and submit thy inventions to his censure.
—*Thomas Fuller, 1642*

The opinions of other people are a stimulus to us, which arouses others (opinions) in us. It is essentially in this way that they (other people) serve us.
—*Claude Bernard, 1850–1860*

... we can learn through criticism of our mistakes and errors, especially through criticism by others, and eventually also through self-criticism.
—*Karl Popper, 1958*

We scientists, and humans generally, are strikingly blind to our own irrationality and errors, even though we can see them clearly in others. The surest way to improve the quality of our ideas and presentation of our work is through criticism from other researchers.

Seek comments on your written work from other scientists before you decide to publish. Do this on your own—not as part of the traditional review process with a journal or publisher. This kind of author-driven review can replace the peer review offered by conventional journals if you choose to publish your work on preprint servers or in post-publication review outlets (see Chapter 32). Publishing on preprint servers also allows you to revise your papers, which means you can also solicit comments after publication and incorporate any changes in updated versions of your work.

Even if you decide to publish your work in a conventional journal, do not expect the comments from journal reviewers to have the quality, detail, and coverage that you are likely to get in comments from reviewers you select. This is so because those from whom you solicit comments have at least a minimal social relationship with you and thus are more invested in giving worthwhile comments than anonymous reviewers who, as gatekeepers, are primed to view you as an inferior or rival. Reviewers you select also are more likely to have the expertise and interest needed to evaluate your work constructively than reviewers selected by editors who may be less committed to the task. I have received many hundreds of reviews from journal reviewers in my career; only a small handful were as helpful as reviews I solicited from others myself. In any event, if you submit your work to a journal, solicit comments on your own beforehand to improve the quality of your manuscript.

Solicit comments not just for completed papers, but also for grant proposals, research plans, and protocols. In fact, comments on your research itself are only useful *before* you actually do the research. The comments that are useful after the research is done relate to interpretation, discussion of other work, speculation, and writing. Reviewers sometimes call for revised or additional analyses after the research is done. Such analyses often are post-hoc exploration (see Chapter 51) and the results may not count as evidence in the same way as those from originally planned analyses.

If you have a problem with an idea you're developing or an aspect of your research, don't wait to get feedback. Instead, discuss the problem with a colleague as soon as you can. Even discussing the problem with a non-scientist can be helpful, as it requires you to describe assumptions and chains of thought, which may let you discover an error. The other person, especially if not trained in your specialty, may also ask questions or give comments that reveal, perhaps indirectly, the source of your problem.

Several kinds of researchers make good reviewers. Colleagues who know your work and background are usually the most likely to give comments. Their feedback is also the most likely to be extensive. It is not important that these colleagues be specialists on your topic or in your field. Criticism may be especially helpful from scientists who do not have the same assumptions and biases you do. Good scientific writing is comprehensible by non-specialists, and comments from such reviewers allow you to assess how well you meet that standard.

Also send your work to those whom you cite and ask for their comments (see Chapter 40). This is a good way to avoid describing their work and ideas inaccurately. Other potential reviewers include scientists whose opinions you value, even if you did not cite them or do not know them personally. I have been pleasantly surprised that some researchers whom I did not know previously have given me comments on my work when I have asked. For many topics, there are few active researchers, and they don't often get a chance to interact directly. Soliciting comments is a great way to initiate this contact and communicate about a shared interest.

Show respect to your reviewers by preparing your manuscript carefully (see Chapter 30), as if it will be published immediately. Assure your reviewers that you want their candid feedback. Praise alone is not useful and does not improve your work. Welcome any kind of comment, from minor editorial matters to critiques of the purpose of your work. If there are particular aspects of your work that concern you, ask your reviewers to give an opinion on them specifically. If reviewers do not respond with comments within the period which they agreed to respond, contact them again and remind them that you are still interested in their comments if they have the time and interest to give them. Remember that they are doing you a favor, so be grateful for any comments that you do receive.

Negative comments, no matter from whom, sting, even though you sought them. Our first instincts (and often, reactions) are to defend, explain, and justify. But it is far better to absorb and consider the comments thoroughly first. Try to adopt the reviewer's perspective as much as possible. Sometimes the reviewer misunderstands your work, perhaps because you wrote poorly. Sometimes you might misunderstand the reviewer's comments. Most reviewers are happy to explain their thoughts further. Ask the reviewers questions, and even debate them on the points at issue if they are willing, thereby permitting you both to clarify different views. Good authors regularly recognize the merits of the reviewers' comments, and good reviewers recognize that sometimes their comments are mistaken.

Comments from reviewers may sometimes touch on matters that you agree are problems, but you may differ on the best way to resolve them. Even after careful consideration, you might still disagree with a particular comment. These differences of opinion are fine. You do not need to take all or even most of a reviewer's advice for it to be helpful. Occasionally, a sharp reviewer notices what appears to be a small problem that, after you examine it further, you realize is a much bigger problem or related to other problems you had not recognized before. Such comments are both scientific revelation and salvation.

Be sure to acknowledge your reviewers in the revision of your work (see Chapter 24) and send them the published version of the manuscript as a small token of thanks.

32

Publish, Then Perish

In vain have you acquired knowledge if you do not impart knowledge to others.
—Deuteronomy Rabbah, c. AD 650–900

Given the knowledge from men in different professions that is not recorded, I am convinced that it exceeds in amount and importance all that which has already been recorded in books, and that the better part of this treasure has yet to be made known.
—Gottfried Wilhelm Leibniz, 1688–1690

Work, Finish, Publish.
—Michael Faraday, 1791–1867

... no work done in the service of investigation is ever lost, not even when carried out under false assumptions.
—Fridtjof Nansen, 1897

"Tell me if anything was ever done," Leonardo da Vinci (1452–1519) wrote in his notebooks, despairing that so many of his projects were never completed. And, ironically, for centuries it was as if even his countless completed projects were never done. Most of Leonardo's scientific observations and insights were in these notebooks, which went unpublished and remained unknown to scientists for more than 300 years. Had his work been published in his lifetime or soon thereafter, the impact on science and technology might have been dramatic.

For modern academic scientists, "publish or perish" is a mantra. Academic researchers who do not publish frequently tend not to receive as much prestige, salary, and other rewards as those who do. Some scientists who do not publish enough might even lose their jobs in research. This dynamic may have little to do with the quality, value, or integrity of the research published. Despite the pressure to publish, scientists inside and outside of academia still do not publish a large fraction of the research they have done. Yet researchers have not truly finished a study until they publish a report on it. To publish is to "make public," and there are many ways to do so, even beyond traditional scientific journals.

The scientific enterprise cannot function without publishing. We scientists learn primarily from the experiences and ideas of others, and whatever knowledge we generate ourselves dies with us unless we transmit it to others. Publishing *all* research is essential not only for this general purpose, but also for reasons that affect us more directly.

Honoring commitments

In most of our research, we scientists incur moral and ethical obligations to complete and report the work. Human research participants often volunteer out of a sense of altruism, believing their participation may produce something of value. A tangible scientific benefit in the form of a published report would seem to be a minimal compensation for disturbing or sacrificing living organisms in research. Similarly, when we disrupt, damage, or destroy irreplaceable or rare materials in research, we owe future researchers, who will not be able to use these resources, a published report. Colleagues and others extend us courtesies and grant us favors that allow us to conduct research, often with the expectation that something worthwhile will come from it. Furthermore, funders, employers, and clients deserve a full reporting of the work they sponsored in good faith.

Preventing publication bias

If you've made up your mind to test a theory, or you want to explain some idea, you should always decide to publish it whichever way it comes out. If we only publish results of a certain kind, we can make the argument look good. We must publish both kinds of results.
—*Richard Feyman, 1974*

Scientists have many reasons for not publishing the results of particular studies. Results that are contrary to authors' hypotheses, as Feynman noted, are a leading excuse. Other invalid justifications include results that lack so-called statistical significance (see Chapter 8), differ from those previously published, are likely to be criticized or unpopular in some way, are difficult to interpret, are old or out of date, are not interesting, and are null or represent failures.

In the conventional journal publishing system, publication is contingent on favorable evaluations by other researchers (so-called peer review). This system underlies, or at least exacerbates, many of the reasons why researchers do not publish their work, as they may give up on publishing a paper after rejection by one or more journals, or never even submit a report at all due to their expectations it will be rejected. Because the reasons for not publishing are related to the results, and not other aspects of the research, the published empirical record on any given topic may become biased, and not just incomplete. By not publishing, scientists actually may tilt collective scientific knowledge away from truth.

Pre-registering your research is essential for keeping a public record of all studies that have at least been attempted (see Chapter 4). Pre-registration also may give us an extra push to complete and report our research, as the registry serves as a scorecard for how well we fulfill our scientific responsibilities.

Establishing priority

Publishing all of your research allows you to establish priority for your discoveries and ideas. It is quite difficult to get credit for work that was never published. The importance of some studies is not always clear until many years after researchers completed them, so avoid regret by publishing all of your work. If your research was important enough to attempt and do, it is most certainly important enough for you to report.

Improving your understanding

Learn as much by writing as by reading.
—*Lord Acton, 1895*

Only through committing your ideas, methods, results, interpretations, and speculation to print can you develop and document them fully. To describe or explain something to others, as in a research report, you must first understand that something yourself. And there is no real way of knowing whether you understand something until you can describe and explain it to others. Therefore, writing up and publishing your research is a crucial step for learning from your own work.

Getting feedback

We do not believe any group of men adequate enough or wise enough to operate without scrutiny or without criticism. We know that the only way to avoid error is to detect it, that the only way to detect it is to be free to enquire. We know that the wages of secrecy are corruption. We know that in secrecy error, undetected, will flourish and subvert.
—*Robert Oppenheimer, 1950*

Sometimes scientists intentionally do not report their research publicly so as to maintain a potential commercial, military, or other advantage for themselves or their sponsors. However, as a long-term approach to science, this strategy is weak. Researchers who restrict their scientific communication deprive themselves of the greatest benefit of publication: feedback. This feedback includes not only criticism, but replication and new research on related questions. Such responses can take time. But if the research is as important as the original researchers believe, others will surely join the discussion and investigation.

Science publishing, liberated

Publishing in conventional journals no longer is necessary for reporting research. Preprint servers and journals that administer peer review *after* publication provide all the functions for good scientific communication. "Preprint" originally meant a version of a paper that had yet to be accepted and published by a journal. However, some researchers now use preprint servers as their primary or only means of publishing their work, the preprint label notwithstanding. Preprint servers allow scientists to return to the generally unhindered mode of science publishing that existed in print for hundreds of years until the early to mid 1900s.

With preprint servers, you can conduct peer review on your own (see Chapter 31) in addition to comments and published responses you may receive on the preprint site or elsewhere. It is easy to produce a paper for a preprint server that looks like an article published in a journal if that is your preference. Publishing on a preprint server is cost-free and fast (sometimes instantaneous, and usually never involves delays longer than a few days). Preprint authors typically have total control of their content, and can update their reports as necessary. Although preprint servers are currently less widely indexed than some journals, you can directly distribute your work to those most likely to be interested in it (see Chapter 40). Regardless, your paper will still be captured in the most comprehensive index of the scientific literature, Google Scholar (see Chapter 5), so will be easily found by anyone who cares to learn about your topic. Moreover, preprint servers offer universal, free access to readers. In short, preprint servers and similar outlets enable free and open scientific communication.

Publishing in journals now is strictly about career advancement and status-seeking by authors or their funders, employers, and clients. Publishing in journals also delays publication, involves a considerable additional amount of authors' labor (and sometimes money), limits readership (for subscription-based journals), and offers authors many kinds of aggravations and disappointments. Peer review often serves as a way to protect dominant views and suppress others, independent of scientific merit. Making publication contingent on the evaluations of others is anti-scientific: although feedback from other scientists is vital (see Chapters 31 and 42), any barrier to the free communication of scientists hampers the development of reliable knowledge. Some researchers get so discouraged by the hostility, pettiness, and censorship common to the peer review process that they quit scientific research altogether.

Nonetheless, authors may still choose or be required to publish in conventional journals. Most journals, though, do allow authors to archive their papers on preprint servers prior to journal submission. At the very least, doing so establishes priority, guarantees your research is entered into the scientific record, and preserves what your work looked like before the influence of journal editors and reviewers.

Some preprint servers even allow publication of conference posters and presentation slides. This is a good option for older projects that you do not have time to write up as full reports. Such posters and slides typically include most or all of the essential details to be worthwhile contributions to the literature. If you no longer are able or willing to complete analysis or write-up for a particular study, you can still fulfill your scientific obligations by archiving the data (see Chapter 9) and registering the study even though you have already collected the data (see Chapter 4).

Even long scientific reports and documents can be published without restriction as self-published books. Many commercial services are available for self-publishing of print and electronic versions of books, including some with no cost to authors.

When planning new research, always reserve a large fraction of the study schedule for reporting (see Chapter 3). Tight schedules can lead to rushed reports, which are more likely to include errors and omissions, or no reports at all. Don't put off writing up and publishing a completed study. Publish while it is still fresh in your mind. There is no value in waiting, as you always have the opportunity to revise in the future. Death could come tomorrow, so publish today.

33

Giving a Scientific Presentation

Don't present a paper—give a presentation.
—Linton C. Freeman, c. 1991

Although spoken presentations have a long history in science, their value is not self-evident. They are poor vehicles for communicating the detailed and complex information in most written scientific reports. Not only are most presentations too short for presenters to give the full information, but their auditory nature contrasts with the visual mode by which most scientists learn—as in reading or direct experience and observation. We learn best when we can digest the information, think, and react at our individual paces, moving forward and backward through the content on a whim and consulting other sources as necessary. Despite this mismatch between presentations and typical modes of learning in science, many scientists still try to make their presentations serve the same functions as a paper. Many of my own presentations have suffered from this fault.

Researchers' *spoken* words in presentations and media interviews (see Chapter 34) do not count much in science. Scientists rely on and cite such statements extremely rarely. Rather, the final words in science are researchers' *written* words —presumably chosen by their authors after much thought.

If a presentation is not and cannot be a paper, then what is its purpose? A presentation is an *advertisement* for a paper or other written account of the research. A presentation is essentially an extended abstract. Many scientists do not read the literature much and rely on presentations at conferences, seminars, and workshops to stay abreast of the scientific action. Thus, despite their defects as a mode for scientific communication, presentations can still serve an important role in disseminating research.

Some scientists also use presentations as a way to display (when presenting) and assess (when listening and viewing) skills and knowledge (see Chapter 36). Presentations can be a kind of scientific audition, especially in the competition for jobs and status. However, in my experience, the relation between quality of presentations and quality of the research ultimately reported in written form (or observed first hand) is modest at best. Some brilliant scientists give weak presentations, and some incompetent researchers are engaging presenters. Don't mistake the cover for the book.

Nonetheless, there are many simple practices we can use to make our presentations effective and truthful advertisements. Even though presentations have been a core scientific activity for centuries, there is very little rigorous empirical evidence on the effectiveness of different presentation strategies. Therefore, my advice is based on common sense and experience, including my own long trail of presentation mistakes. Beyond the basics I cover, use whatever approaches to presenting that seem best for you and your specific situation.

Preparation

Reading a paper word for word as your presentation is almost guaranteed to annoy or bore your audience. As Freeman emphasized, audiences expect a presenter to be active. Text meant to be read silently—as in most scientific writing—tends to be difficult for listeners to follow when read aloud. But if you don't read your paper, how can you know what to say? Presenters who are nervous or lack confidence often want something to help them remember what to say because the anxiety of speaking to an audience can be overwhelming. But even confident, relaxed presenters need a way to ensure they say everything they intend to say.

Notes on cards, paper, or presentation software are an effective safety net for remembering what to say. The notes can range from very brief one- or two-word topics to cover to scripts for the full presentation. Scripts are planned *spoken* words, not excerpts from a written paper. Reading a script prevents making much eye contact with the audience, which is a major downside. However, I have seen researchers, especially those presenting in a second language, give decent presentations with this approach. Less detailed notes allow a presenter to act more naturally and give more eye contact to the audience.

Visual aids, such as slides, overhead transparencies, or handouts, can also function as a sort of notes or memory cues for what to say. At the other end of the spectrum, some scientists don't need concrete reminders. They may memorize an outline and improvise their precise words as the presentation unfolds. A few researchers even memorize their whole presentations, including words, intonations, and actions.

Regardless of which approach you use, practice is essential for every presentation, no matter your level of experience or comfort with public speaking. Practice giving your presentation, speaking out loud, by yourself or with others at least once and ideally more. Time your practice presentations, and cut material as necessary to ensure it runs less than your allotted time. Solicit feedback from anyone who listens to your practice presentation and revise as you think best. Also, identify the parts of your presentation that you would skip or condense if you run short on time, for whatever reason. A planned retreat is better than a panicked one. It is frustrating to watch presenters jumble their presentations when they realize they don't have enough time to cover everything they had planned.

Scout the physical set-up for your presentation and adjust to any constraints. For example, if you tend to walk when you are presenting, but will speak from a lectern or a fixed-location microphone at your next presentation, practice your presentation while standing still. Arrive as early as possible at your presentation site to get everything you need ready, including testing any technical equipment (computers, projectors, microphones, lights, etc.). Audience members do not like to wait for presenters to fix problems that could have been resolved before the presentation.

I believe that with presentation experience comes greater ease. I was terrified of public speaking when I began making presentations, and my nervousness showed for years. My fears and tension have gradually faded over time, and consistent preparation has been one reason for this improvement.

Manner

When you are presenting, be yourself. If you try to act in ways that don't come naturally to you, the audience might regard you as fake, which detracts from your advertisement. Similarly, acting unnaturally increases your stress, which may make it harder for you to present well. Therefore, use humor, expressive enthusiasm, gestures, or movement in your presentations only if you normally use them in interacting with scientists in other contexts.

Scientific audiences, especially at conferences, are linguistically and culturally diverse. Therefore, avoid idioms and colloquialisms in presentations, because such words and phrases will be empty or confusing to members of the audience who are unfamiliar with them.

Structure

A research report has a standard structure and conventions (see Chapters 19–23). Presentations are different. There are diverse ways to organize an effective presentation.

Some researchers advocate following a fixed structure. One such format is to state the main points of the talk, then describe those points and the evidence that supports them, and finally review the main points again ("tell them what you will say, say it, and then tell them what you said"). While this may be a reasonable approach for long presentations (say, more than 30 minutes) or classroom lectures, it is very redundant (see Chapter 29) and formulaic. Presentations with this format, even if otherwise effectively delivered, tend to be monotonous, particularly in conference settings with many such presentations in succession. If your presentation will last less than approximately 30 minutes, use a different approach. Although some readers of a written report may skip to its concluding summary, your presentation audience will likely have heard your whole presentation, making a summary (of a summary!) unnecessary. It is enough to introduce your topic, outline your methods and results, and offer some comments and concluding remarks.

Also consider taking a more creative approach that might especially hold an audience's interest. I have seen some presentations given essentially as stories, others beginning with and focusing primarily on one or a few images or animations from the research, and still others that begin with the presenter engaging the audience in a task (or simulation of one) central to his or her research (such as related to observation, or self- or thought-experimentation). The potential variety might be limitless. Taking one of these alternate approaches requires careful thought, making sure that it fits your topic, circumstances, and audience.

Simplicity is just as important for presentations as it is in all phases of scientific research and writing (see Chapters 13 and 29). Spoken presentations cannot contain much information compared to written reports because speech is much slower than silent reading (see next section on Speaking). Particularly at conferences, audience members appreciate streamlined presentations that focus on key themes without complex details.

Speaking

Speak briefly. Just as with scientific writing (see Chapter 29), say what you need to say and no more. It is better to say too little than too much; avoid boring or irritating your audience. During the question and answer session or after the presentation, audience members can ask for more information if what you said was too little. In the thousands of presentations I have attended, I have never heard an audience member complain about a presentation that was too short. However, presenters who keep their presentations brief earn praise.

Speak loudly. Very few presenters, in my experience, are ever too loud, but many are too quiet. If you don't speak often in front of groups, the appropriate volume may seem too loud, even though it is not. If you are uncertain about how loud you should speak, ask your audience at the beginning of your presentation whether everyone can hear you well and change your volume accordingly. Especially if you are speaking without a microphone, it is difficult to speak too loudly.

Speak slowly. I recommend a rate of about 100 to 120 words per minute, which is slower than regular conversation. Members of the audience are in learning mode, so fast speech means too much information for them to process effectively. My recommended rate translates to about 1,000–1,200 words in ten minutes, 2,000–2,400 minutes in twenty minutes, and 6,000–7,200 words in a full hour. The scientific content of your speech might even be less than these counts. Your rate may be reduced by your adjusting to audience reactions, working with visual aids, adding unnecessary words unintentionally (e.g., umm, you know, OK, etc.), moving about, taking a drink, and other actions. Conference presentations, therefore, include fewer words than the typical length research report in most fields. This underlines the fact that most presentations can really only be advertisements. Most scientists are excited or nervous before and at the start of their presentations, which can lead to speaking too fast. Speak more slowly than you think you should. You can estimate your speaking rate when you practice by recording and then replaying a sample of your practice presentation.

Speak directly to your audience. If possible, stand in front of and face your audience. Limit the time you spend looking at (and talking toward) visual aids (such as projection screens) or notes. Although we scientists learn primarily by reading, we humans are especially attuned to learning through face-to-face contact. For some presenters, direct eye contact with members of the audience is distracting or intimidating. The solution in this case is to look at audience members' foreheads or between audience members (if the audience is large and deep enough), so that it looks as if you are making eye contact even when you are not.

Visual aids

I have attended many excellent presentations. In some, the presenters used no visual aids, while in others, the presenters used multiple media extensively.

In my opinion, there are five reasons to use visual aids in a presentation. First, in many fields, scientists *expect* visual aids to accompany every presentation. If you think visual aids are not necessary for your presentation, override such norms. Second, some presenters may be self-conscious, and visual aids focus the audience's attention away from the presenter, thereby reducing his or her anxiety. Similarly, some presenters may worry that they will not speak coherently and use visual aids with lots of text as a back-up form of communication. Avoid using visual aids as a diversion tactic or fail-safe. If this crutch is necessary for you to be able to present, use it, but try to reduce your dependence on it. Fourth, visual aids can serve as an outline or a set of memory cues for the presenter. If this is the only purpose of the aids, consider another way to remind yourself of the presentation material, as visual aids are for the audience, not the presenter. Fifth, visual aids may be the only way to communicate some content well, such as mathematical formulas, tables, graphs, maps, models, photographs, drawings, symbols, animations, or quoted text for detailed analysis. This is the best reason to use visual aids.

Because most of us scientists were trained in academic settings and many teach in them, we often treat presentations like class lectures, with text-oriented visual aids to be recorded by members of the audience as notes for later study. But those who attend scientific presentations rarely take notes on presentations, much less study them. The experimental evidence is very inconsistent about the merit of including text-oriented visual aids that are redundant with a presenter's spoken words, and none of it is based on presentations to researchers. Therefore, I suggest keeping visual aids (or parts of them) that have text only to an absolute minimum. The main exceptions to this advice are when you present to an audience that likely includes persons who do not understand your accent, do not understand speech in your language well, or are deaf or hearing-impaired. If you present to an audience that includes persons who cannot *read* your language well, make hand-outs with essential text in their language, if you are proficient in it.

When preparing your visual aids, make the smallest relevant details large enough that they can be seen easily by those who have poor vision (most researchers are middle-aged or older) or are located at the back of the presentation room. Usually, this means that images, numbers, and text appear huge, for example, in slides or transparencies as you prepare them.

Just as with your spoken words, make your visual aids few, clear, and simple (see Chapter 25). Give your audience time enough to absorb each image, which is typically longer than it takes for you to discuss it. Rapid fire images, like fast speech, give the audience no chance to digest the content. A scientific presentation is different from presentations in which the presenter seeks to motivate, persuade, entertain, or sell. Flashy or emotional approaches hinder rational, scientific communication. Similarly, avoid gimmicks that distract and lack scientific meaning, such as irrelevant image templates and borders, complex color schemes, and visual transition effects when changing slides.

Props, such as reproductions of relevant specimens or tools used in your research, are unusual but interesting visual aids. If they are cheap to make, you can also give them as handouts. They may make your presentation more interactive and are lasting physical reminders of your presentation.

Question and answer period

Beyond advertising your research, the other main scientific value of giving a presentation is to receive critical feedback. When you prepare your presentation, anticipate likely challenges to your work. Plan responses to those criticisms that you can imagine.

Immediately after the main part of your presentation and before the question and answer period, shift mentally from speaking to listening and thinking. Most presenters ride a surge of adrenaline during their talks, and it takes conscious effort to change modes quickly and effectively.

If you are not certain about the intent of an audience member's question, try to restate it in terms that make sense to you and ask the questioner if that is what he or she meant. It is better to clarify first than waste time answering the wrong question.

It is fine to admit that you don't know an answer to a question or that you need some time, after the question and answer session, to consider a criticism or question before you could give a response. Write notes on the questions you get, especially for those to which you don't have any good responses yet. Do this before you leave the presentation room, as it is easy to forget such valuable challenges.

Focus on the scientific content of a question, not the questioner's emotion. Rarely, a questioner will be openly hostile to a presenter (often the hostile questioner is higher status than the presenter, and feels threatened by the presenter). The key is not to respond with an emotional reaction. Attempt to convert exchanges during the question and answer period, including awkward ones, to lengthier, more relaxed private discussions after the presentation.

Despite the potential value of question and answer periods, keep your expectations low. Presenters and questioners respond to each other with little time to reflect on matters they often have not thought about before. Questioners don't have the full information on the presenter's research given the necessarily incomplete and abbreviated material in the presentation. And sometimes presenters get no questions at all. You cannot control the interests, knowledge, attitudes, or size of your audience.

After the presentation

Always be prepared and willing to share any visual aids or spoken information (translated to written form) from your presentation on request, even for preliminary work. A presentation is a kind of publication—you have made your work public, and therefore are obligated ethically to share what you presented with others, including those who did not attend your presentation. And if you have already written the paper on which your presentation is based, tell the audience how they can get an electronic copy or bring paper copies for those who are interested.

34

Dealing with the Press

Nothing can now be believed which is seen in a newspaper. Truth itself becomes suspicious by being put into that polluted vehicle. The real extent of this state of misinformation is known only to those who are in situations to confront facts within their knowledge with the lies of the day ....
—Thomas Jefferson, 1807

If you're representing yourself as a scientist, then you should explain to the layman what you're doing—and if they don't want to support you under those circumstances, then that's their decision.
—Richard P. Feynman, 1974

I don't like doing interviews. There is always the problem of being misquoted or, what's even worse, of being quoted exactly, and having to see what you've said in print ... I think Nabokov may have had the right approach to interviews. He would only agree to write down the answers and then send them on to the interviewer who would then write the questions.
—Stanley Kubrick, 1982

Most researchers feel flattered and excited when journalists contact them to talk about their work. Many scientists want to communicate with the press in the name of educating the public, a noble purpose. However, non-scientific and more ignoble motives may also be involved. Press attention appeals to our vanity. By discussing their work with the media, some researchers also hope to advance political or ideological agendas, boost support for government research funding, or attract private investment in associated commercial ventures.

Risks

Journalists come calling only when they sense that your work might be translatable into an interesting story. Your research may be only part of the story. The reporter determines what the story is and how to tell it. Most likely, journalists do not contact you because of the quality or creativity of your work. Given their typical lack of scientific training, journalists generally are unable to judge such matters.

Journalists do not necessarily want to educate the public. In crafting the story, they may reframe, overstate, distort, or sensationalize your work in their quest to attract readers, listeners, and viewers. Some reporters may be sloppy, and misquote you, or quote you out of context. Still other journalists may use you as a foil and aim to portray you and your work unfavorably. There are fair, competent, and ethical journalists covering science who want to report your work accurately, but it is often difficult to determine whom they are at the outset. Moreover, journalists with good intent may not be responsible for the final coverage. Editors, space limitations, and deadlines, for example, can influence strongly how the story ultimately appears.

When you are interviewed, there is also a risk that you will misspeak or say something that you regret but be quoted accurately, as Kubrick feared. It is difficult for most researchers —or really anyone—to be perfectly articulate on the spot in response to unexpected questions. We are usually lax in impromptu communication, yet the media report may capture what we said verbatim. When the account is in print, our spoken comments may have an impact as if we had written them. An interview with a journalist can involve a scientist repeating or using words that the journalist uses but that the scientist wouldn't otherwise normally use, because it's a conversation, after all.

As an interviewee, you have no control of how your work will be reported. I have been interviewed many scores of times by journalists, but fewer than a handful honored their promises to let me check the accuracy of their planned stories. Only a similar small number even fulfilled their pledges to share the stories once they were published or aired.

Strategies to promote accurate coverage

If you choose to interact with the media, engage journalists proactively to increase the chance they cover your work correctly. Well-designed press releases can help prevent some of the problems that stem from interviews. Many reporters create science stories entirely from press releases, or take from other reporters' coverage of the same story. Press offices for universities and other organizations often are eager to promote their researchers' work by creating and distributing press releases. However, press office staff also tend to have little scientific training. Fundamentally, their jobs are to market the institution, and reporting research accurately is not necessarily their highest priority. Therefore, it is *your* responsibility to control the content of the press release. Ensure that the press release is not made public before you review it. Do not approve anything that misrepresents or oversells your research.

If you do give an interview, plan beforehand a very short (2–3 sentence) non-technical summary of your work in lay terms. Write it down or memorize it, so that you can give it readily. Also compose and rehearse answers to other likely questions. Consider recording the audio of an interview or conducting it in writing, such as by email. This makes clear to the reporter that you have an independent record of what you said. You can use it to correct mistakes the reporter makes, albeit after the fact, if the publication or media channel allows corrections. Keep in mind that any damage done by the original story likely will remain, as corrections gain almost no attention. If a reporter misrepresents you or your work severely, you can submit a letter to the editor or an opinion editorial to rectify the record thoroughly. As with corrections, such responses may not even be options or may be missed by most of those who read, saw, or listened to the original story.

If you comment on others' work or ideas, treat the opportunity as you would a direct debate with other researchers (see Chapter 43). Stick to rational and objective points, and avoid attacking others' characters, impugning their motivations, or using other irrational and irrelevant tactics to denigrate other researchers and their ideas. Reporters often like to highlight criticisms between researchers, even (or especially) if those they quote make emotional and immaterial comments.

Alternatives for educating the public

There are other ways that scientists can educate the public that may be better and more efficient than through the press. Take control of the message and the medium. For example, write columns, magazine articles, blogs, and books; produce videos and audio broadcasts; and give public lectures geared to your intended audience. With such approaches, you can ensure the integrity and context of the information and actually increase your audience's understanding.

Social Sphere

35

Working with Other Researchers

Too many cooks spoil the broth.
—*English proverb*

... the leader of a team needs to be actively engaged on the problem himself.
—*W. I. B. Beveridge, 1950*

Doing research with other scientists can increase the fun of science. The most enjoyable projects I have done were enjoyable because I shared the struggles and thrills of research together with colleagues and mentors. Collaborative projects also let you learn from or mentor other researchers. Perhaps the most natural reason to work with other researchers is to do a project that you could not do or do as well or as fast on your own. Other scientists' expertise can complement your own, and working with other researchers can also give you access to data, equipment, materials, funding, study sites, and populations that would otherwise be inaccessible to you. Despite these advantages, working with other researchers involves significant challenges and risks.

Team size

Research teams have been growing steadily larger, on average, for many decades. Part of this growth is due to the increasing complexity of research and the corresponding need for different specialists in different aspects of the work. However, this growth in team size also is due to non-scientific reasons. Researchers often join together in teams of supposed experts as they seek competitive funding. In such settings, small teams may *appear* less competent. Scientists also sometimes build teams of researchers with good reputations, even if their skills and knowledge are redundant, hoping to improve their chances of funding and publishing in high status journals. In addition, some researchers have assistants do what they regard as low status work, such as data collection and data management, and, sometimes, even conceptualization, study design, data analysis, and writing—effectively all research itself. Furthermore, when scientists do too many projects, they often no longer have the time or ability to do the research themselves even if they wanted to, and must find others to do it for them (see Chapter 38).

Large teams come with many costs. Decreased independence of thought is one major drawback. As with all groups, teams of researchers tend to involve social pressures. Lower status individuals defer to higher status individuals, those in the minority on some matter defer to the majority, and team members avoid disagreements to promote social harmony and maintain their income (the latter in funded projects for team members who are not project leaders). These pressures grow with team size and threaten the reasoning (see Chapter 56) and doubting (see Chapter 61) that are essential to good scientific research.

Inefficiency also grows with team size. Scheduling meetings and other joint activities often is difficult for teams with more than a few researchers and typically causes delays. Large teams tend to include researchers who work on many different projects and have, as a result, short attention spans and limited capacities to fulfill their duties on any one project. When there are several main intellectual leaders on a project, attempts to balance divergent opinions can stall progress, introduce inconsistencies, or reduce the research to blandness.

Moreover, large teams tend to divide the labor by giving different tasks to different members. This diffuses responsibility and inhibits project leaders' abilities to control quality.

Therefore, avoid creating large teams unless absolutely necessary for scientific purposes. To preserve as much benefit and reduce the drawbacks of collaborations, I recommend having no more than three primary scientists or leaders on a project, and no more than five researchers active continuously from start to end of the project. There can be almost an unlimited number of researchers who perform narrow, focused roles who do not interact much or even know most others involved with the project, albeit with the costs from the increased division of labor. Beyond these limits, in my experience, a phase change occurs in the group dynamics, with interactions becoming bureaucratic—formal, hierarchical, inflexible, and committee-like. For most projects, it is very difficult to manage practically and socially with more than three researchers who will make the major decisions on design, methods, and interpretation.

If collaboration is necessary, the optimal team size may be two or three researchers *in total*, everything else equal. Ideally, these team members have overlapping expertise and project duties so that a main benefit of collaboration—mutual criticism—is possible and meaningful. Of course, small teams may lack skills and knowledge for some aspects of a project. The best way to handle these gaps is for team members themselves to learn whatever is needed and solicit feedback from experts on project plans, results, and interpretations (see Chapters 31 and 40).

Choosing colleagues

The first step in developing positive relationships with collaborators is to choose them carefully (see Chapter 36). In my experience and observations, the most productive and fun scientific collaborations occur when team members like, respect, and know each other well. Solid relationships allow collaborators to negotiate responsibilities, differences of opinion, money matters, and authorship effectively, and ensure that work gets done properly and on time. Such relationships also allow team members to be comfortable making mistakes in front of and challenging each other. Many long-term research partnerships are like healthy marital or family relationships built on trust, fairness, and friendship.

Collaborations among researchers who have no prior relationship with each other still can be very positive and constructive. I count many trusted colleagues whom I did not know prior to working together on a research project. Nonetheless, such partnerships carry a higher risk of bad outcomes, so adjust your expectations of success at the outset accordingly. Nearly all of the projects in which I was involved that went poorly involved researchers with whom I had no prior relationship.

Of course, in many research settings, scientists are not able to choose with whom they work. Teams may be determined by an organization's formal structure, including hierarchical relationships, or by a project leader's or client's choices without your input. In such circumstances, it nonetheless pays to *understand* your colleagues as best as you can.

It is essential to know your colleagues' motivations for participating in a joint project. Their views on the primary purpose of the research—career advancement or scientific pursuit of truth—indicate their likely behavior on the project. Differences in such views among team members can cause much conflict. These two purposes are not necessarily mutually exclusive, but usually one is more important to a researcher than the other. Many researchers might not be able to admit to others or themselves that career advancement has anything to do with their motivations. Some signs of a dominant career motivation include belief that only certain high status publication outlets are acceptable; avoidance of some ideas, topics, or interpretations because they may be too controversial among their peers (no matter how appropriate for the research at hand); and concern about the status (as opposed to expertise) of other collaborators.

Some career-oriented researchers seek nominal collaborators of high status whom they expect will contribute little to no labor or expertise to the project. Such additions to the team roster boost project leaders' reputations (and that of the project) by association, increase chances of grant/contract funding and publication, and help project leaders acquire influential professional allies.

In other cases, researchers seek high status or expert scientists as collaborators, expecting sincerely that they will contribute meaningfully to their projects. Such individuals tend to be high-risk collaborators, as they may have insufficient time (or expertise—see Chapter 38) to do the work for the project well or at all.

As a project leader, decide what you are willing to offer potential collaborators (authorship/author position, payment, role, and degree of control over design, methods, and interpretation) *before* you invite them to join the project. If another scientist invites you to join a project, discuss these matters explicitly and reach terms acceptable to you before agreeing to join.

As with all cooperative ventures, working with other researchers risks exploitation. Watch out for idea bandits. They are typically ambitious researchers who rely on others for research project ideas to execute. Idea bandits tend to be skilled socially as they charm more creative scientists with deference, enticing them to share plans, hypotheses, and advice. Such seduction may even go as far as inviting the creative researchers to write critical sections of research proposals. Once such projects begin, however, the idea bandits have already obtained what they need, and may end their scientific relationships with the creative researchers. Idea bandits rarely give credit to their sources of ideas, so they might appear as creative themselves. For subsequent projects, idea bandits simply target other creative researchers for ideas, forming a trail of scientists whose goodwill has been abused.

Some idea bandits use a different approach. They recruit inquisitive students and staff, and harvest their ideas. The exploited researchers in such cases usually get at least some, but less than deserved, credit for their contributions.

Standards of professionalism, prior verbal agreements, job descriptions and employment policies, and contract terms may indicate expectations for the different researchers in a joint project. However, they are often violated. Researchers who feel wronged by a collaborator in a project gone awry may, in principle, have legal or other recourse to seek a remedy. In practice, though, such responses are usually unproductive, as they may involve substantial money, time, stress, and reputational risk. The best insurance against a soured collaboration is a good prior relationship with your colleagues.

Leading projects

Before building a team for a project you will lead, first consider your own strengths, weaknesses, and capacity to do the work. By knowing your deficits and needs, you can focus on finding others who have the necessary skills, knowledge, and time. If you can fill a role well enough yourself, do it—don't delegate it to someone else.

Select team members and assign their roles carefully. Scientists vary tremendously in their strengths and weaknesses, just like other people. Some scientists excel at thinking—generating ideas, speculating, interpreting, etc.—but struggle with doing—starting and finishing tasks. Brilliant scientists are of little value if you are counting on them to do things they are unlikely to complete. Some researchers may do tasks, but perform them poorly. Other researchers may do good research on their own, individually, but do not work well with others, or cannot effectively give or receive direction. And still other scientists may neither think nor do, but just ride on the achievements of their students, staff, consultants, and colleagues (most common among high-status researchers; see Chapter 36). If you are responsible for a project or part of one, be prepared to add or drop researchers from the team, or restrict or expand their duties, to address deficits in skills and work habits. The success of a collaborative project may depend on such actions.

Research managers often are project leaders, but not engaged in the research themselves. In such situations, the best approach is for the manager to delegate leadership to a scientist who is directly involved in the research. Many research managers cause their projects to fail or perform poorly because they give insufficient attention to projects or no longer have the time, skills, knowledge, and experience needed to run them effectively.

One of the worst times to add a team member is near the end of the project, in the interpretation stage after the team has obtained results. Team members who worked hard from the beginning might resent the addition who will get credit based on much less work. Project leaders can anticipate the help needed throughout the project by planning the study (see Chapter 3) and drafting the pre-registered study protocol (see Chapter 4) themselves. If unforeseeable events occur during the project that do require adding a team member, consult with your team members before doing so.

Have members give regular feedback on each other's ideas, plans, and results, including your own. Such feedback is valuable even if those offering feedback are not familiar with the topic at hand. In presenting to fellow team members, researchers are forced to describe their work in clear, simple terms. The feedback process may also allow non-expert team members to identify and criticize implicit assumptions.

Defining roles

For an efficient project, it is essential to define team members' roles clearly and inform members of each person's responsibilities. Without such explicit statements, members may duplicate each other's work, do irrelevant work, or neglect important tasks. This scenario is particularly common when team members write, edit, and revise a manuscript without guidelines and a coordinated sequence for their input.

Communicating with colleagues

Regular face-to-face contact with your collaborators is ideal, but not always possible. Many joint projects nowadays involve team members who are located far from each other. Communication by telephone or Internet voice/video calling can be useful in such circumstances, but there are limitations. If you and your collaborators do not share the same first language or accent, it may be difficult to understand each other's speech without the benefit of nonverbal communication or the comfort of conversation in person. Common technical problems with these communication modes, such as poor audio quality, noisy signals, dropped connections, lack of simultaneous two-way transmission, and misalignment between eyes of participants and video cameras also interfere with clear and effective communication. Thus, written communication, whether by email, postal letter, or some type of online platform may often be the main or even only mode of communicating for some teams. For example, I have had many colleagues in other countries with whom I have worked, separately. None of them spoke English as a first language. My contact with each of them was entirely by email and instant messaging.

When meeting in person, by phone, or via Internet voice/video calling, make sure a scientist familiar with all the technical matters takes good notes on what team members discussed, including all decisions made and reasons for them. Notes are essential even if the meeting involved only two team members. Distribute the notes to the whole team so that members can coordinate their work.

Handling disagreements

A basic challenge in joint projects is getting a set of individuals with different ideas and preferences to agree on a single course of action. Ultimately, to publish results, the team must, with few exceptions, present the project in a united, coherent way. At the extreme, conflict between team members can disrupt, even kill, projects and erode team members' mental health.

Many projects involve working with other scientists trained in different fields as well as non-scientists (for example, clinicians, engineers, technicians, administrators, programmers, writers, etc.). Team members may vary in the settings in which they work, such as academia, government, industry, or other contexts. Team members may also vary in their theoretical orientations, preferred methods, and interpretations of results. You may be the only researcher from your field, or even general branch of science, on the project. This diversity in backgrounds allows project members to learn new ideas and approaches, but also may lead to clashes in perspectives and goals. To reduce the chance of unproductive arguments, stay focused on the needs of the project and leave aside differences with others that are not related to the project. Just as in interactions involving others who have different cultural backgrounds than your own, listen with patience and ask questions respectfully. Try to understand how your project colleagues view problems and solutions to them. Present your own ideas as simply as possible and explain any technical jargon you use. Emphasize points of agreement and shared understandings with other team members to reinforce your common purpose of advancing the project. This helps clarify which matters the project team must discuss or work on to move forward.

Differences of opinion between researchers on a project are common regardless of their backgrounds. If you are a full collaborator in all or most aspects of the project, you may argue for your ideas with your colleagues and try to reach a mutually satisfactory resolution. If, however, you agreed to participate in the project with a defined limited role, raise questions and suggestions outside that scope carefully—but do raise them if they pertain to key aspects of the project. Beyond this, pushing your views on topics outside your specific role is likely to annoy your colleagues. If you have a fundamental disagreement on a matter important to you, decide whether you can accept your colleagues' different idea or approach just for this project. Graciously excuse yourself from the project if you cannot accept what they want to do.

Many project teams begin their work eagerly without planning the project comprehensively and carefully. Such hasty starts increase the chance of disputes among team members later in the project. By developing thorough and detailed pre-registered study protocols with your colleagues, you can prevent many disagreements and conflicts that would otherwise arise.

Sometimes, colleagues may have similar views and like each other, but find it difficult to work together. One may try to force his or her will on the other, their work styles may clash, or they argue with each other often. If these problems persist over long periods with no hope for resolution, it is probably best that they do not work actively together any further. Such colleagues can still discuss each other's work and benefit from the feedback—perhaps even more so after they are no longer confined in a joint project.

In any collaboration, expect disagreements, and be worried if they do not occur. A healthy collaboration involves open communication and free sharing of views. No two researchers have the same ideas, even on small matters. If disagreements do not arise, it may mean that some collaborators do not feel free to speak up. Disagreements can be very productive, especially if they highlight prejudices and unfounded biases, prompt attempts to evaluate competing ideas empirically or develop better ones, or moderate views when there is no empirical or rational basis for favoring one or another.

36

Recognizing Scientific Skills,
Knowledge, and Character

*For authority proceeds from true reason, but reason certainly does
not proceed from authority.*
—Johannes Scotus Eriugena, A.D. 864–866

They looke into the beauty of thy mind,
And that in guesse they measure by thy deeds.
—William Shakespeare, 1609

*At 100, I have a mind that is superior—thanks to experience—than
when I was 20.*
—Rita Levi-Montalchini, 2009

To learn how to do scientific research well, it is essential to
identify others with high levels of skill, knowledge, and
character.

Character is especially crucial. Mentors, including teachers
and advisors, who have your best interests at heart are more
likely to be useful to you than those who see you as only a tool
for their own purposes.

Character is equally important when conducting research
with others. Collaborative work involves shared risk: multiple
researchers give their time and effort to a project that can fail if
any does not fulfill his or her promised contribution.
Collaborators who complete their work, but without integrity,
damage the cooperative venture even more than if they had not
done their work. The risk of collaborative work also often is not
shared equally. High status researchers and those in roles with
financial and legal control, such as principal investigators on
grant-funded research, can decide to accept the work of their
collaborators but deny them previously agreed-upon rights and
rewards, such as intellectual input, authorship, and payment.

Thus, the character of your collaborators has direct and personal consequences for you.

Poor indicators

Conventional criteria that scientists typically use in assessing other scientists are, in my experience, not useful. Academically trained scientists, clinicians, and other professionals tend to focus on degrees, titles, institutional affiliations, academic lineages, awards, and counts of publications and citations when evaluating others.

Although these attributes probably have some relationship to skills and knowledge, they are far from perfect indicators. Some of the most talented scientists I know have little formal scientific training. And I have encountered many scientists trained at elite universities who are unable to perform basic research tasks or reason coherently on scientific matters. Of course, most scientists before 1900 also lacked advanced degrees, and many were not formally trained at all. Similarly, powerful positions in universities or other organizations also are not good indicators of scientific attributes, as they often reflect individuals' political and administrative skills, not scientific talent.

Publications alone also are not a reliable guide to a researcher's skills and knowledge. Some scientists depend heavily on unacknowledged others to do parts of or all of their research (such as students, assistants, ghost writers, or consultants). Researchers with very high publication rates are very likely to be research managers (see Chapter 38) or others who have tangential or superficial involvement in the research reported.

The number of students and/or collaborators a researcher has is not a good indicator of skills, knowledge, or character. A large number could mean that the scientist is a valuable teacher or collaborator, or that the scientist is a parasite, living off the contributions of others. High status researchers in particular tend to recruit and attract able students and collaborators, and it can be difficult to discern whether the mentors, proteges, or both are skillful and knowledgeable.

Even scientists' records as principal investigators on research grants are not good indicators. Just like some apparent mentors, many principal investigators are figureheads who supervise and get credit for the work of their colleagues, staff, and students.

Useful indicators

In evaluating potential mentors, employees, collaborators, and sources of information and advice, look beyond status markers and assess skills, knowledge, and character directly. Admittedly, this takes some effort and time, but the stakes make the investment worthwhile.

Perhaps the best way to evaluate scientists is to observe them in the context of their daily work. Supplement your own observations with the views of others who have worked with or near a particular scientist, in whatever capacity. Make sure to consider others' reports of the scientist's behavior, and not just their interpretations of it. By noticing whom other researchers seek for help and advice, you can learn who the most valuable scientists are, sometimes quite to the contrary of traditional markers. For instance, in the early 1900s at the Woods Hole Marine Biology Laboratory, Ernest Everett Just was the unofficial sage whom other scientists constantly sought, so much so that he often had little time for his own research. He served unselfishly in this role, despite his otherwise low professional and social status and the presence of many other distinguished researchers.

The content of researchers' publications may give glimpses of a researcher's scientific character from afar. In the acknowledgments, does the author recognize others for comments and persons who undoubtedly must have helped to enable the project's success (see Chapter 24)? Does she or he follow good scientific practices, such as pre-registering research, archiving data, questioning and doubting popular views publicly, evaluating and rejecting their own hypotheses, testing rival hypotheses, using others' critical feedback as inspiration for new research, seeking and considering all relevant evidence, offering hypotheses and research questions in his or her writings (intellectual generosity), and other practices highlighted in this book? The acknowledgments in students' theses and dissertations may be very useful in assessing the skills, knowledge, and character of professors and others in graduate training programs. Whom do students recognize for help (and for what kinds of help) and how do their levels of gratitude vary among those they acknowledge?

Pay close attention to any description of how each author contributed to a published work. Many researchers gain co-authorship for trivial actions, especially if they are high status or are scientific managers. Keep in mind that these descriptions of author contributions may exaggerate the roles of some authors, especially those in more powerful positions. Look for scientists who usually are actively involved in research—hands-on participation in study design, data collection, data management, and data analysis—rather than just supervision.

It can be revealing to compare your own observations of researchers with inferences you make from publications. For example, contrast their published writings with their informal communication with you by email, phone, or in person. Sometimes, there is a large gap between these sources in a scientist's apparent skills, knowledge, and language capability, implying that someone else must be responsible for key parts of their work and success.

With other factors held equal, older active scientists may be especially valuable and useful mentors, advisors, and collaborators. Mere technical knowledge and skill can be useless when applied without sound judgment. Wisdom usually grows with experience, and senior scientists tend to have more research experience than junior scientists.

37

Working with Research Partners

Everything in the world we want to do or get done, we must do with and through people.
—Earl Nightingale, 1966

Many scientifically sound projects fail entirely because project leaders do not develop and maintain good relationships with individuals and organizations outside of their project teams who are essential for conducting the research. Such research partners include gatekeepers, hosts, and support staff at field sites; vendors or providers of equipment, materials, and services for the project; recruiters and referrers of study participants; and administrators in your and your colleagues' organizations, among other types of project facilitators.

You may think your project serves a noble and important purpose, but most non-scientists, including your partners, may not. As a scientist, you are not entitled to special consideration, donations, or discounts. To your partners, you are often just another tenant, customer, or burden.

Many scientists experience a kind of culture shock when working with research partners. The real world does not run on ideas, academic degrees, and scientific prestige, but on money, political power, personal connections, laws, regulations, and bureaucracy.

Establish or re-establish relationships with your partners in the early stages of designing and planning your project, especially for field research. A potential project is feasible only with good relationships in place. Formalize these arrangements for project activities, ideally with a letter of support, memorandum of understanding, contract, binding quote, or some other kind of written agreement.

Until your project starts, stay in regular contact with your partners so they remember your upcoming work together and to ensure that they will still be available. This communication also allows you to learn of new developments that might affect your plans. Often with little notice, companies and other organizations can move locations, change their products and services, raise their prices, lose staff (including your contacts), or close. Freelancers with whom you plan to work may also move, change careers, or otherwise become unavailable to your project.

Keep communicating with your partners frequently once your project begins if your routine work does not involve direct contact. Get to know partners personally who play critical roles on your project; treat them as equals and team members. Cultivate support for and interest in your project by giving updates and feedback tailored to your partners, especially on matters relevant to them. This is also a simple way to show appreciation for and respect to your partners.

Bureaucracy can complicate a partnership. Usually it is not enough to have relationships with just the other persons with whom you would work directly. You may also need to have relationships with administrators, including managers and executives, and possibly boards and committees. Such persons and groups tend to be difficult to contact, have little time to deal with you, and understand research poorly. This means that developing or modifying a partnership can take a while.

38

Managing vs. Doing Research

... I have never had any student or pupil under me to aid me with assistance; but have always prepared and made my experiments with my own hands, working & thinking at the same time.
—*Michael Faraday, 1857*

I like being the person who not only thinks of scientific questions, but also performs the experiments. I don't want to miss the eureka moments ....
—*Irini Topalidou, 2018*

Most professional scientists are generally involved in doing, teaching, or managing research. Doing research means conceiving and designing projects; collecting, managing, and analyzing data; and/or reporting research in written or spoken form. Teaching research naturally means instructing others in these activities, and implies that the teacher has the specific skills he or she imparts. And managing research means supervising or overseeing others who do the actual research.

In the past, most scientists did and/or taught research. Now, research projects in many fields have become complex enterprises involving many scientists and other staff. Some scientists, especially as they progress in their careers, end up managing research done by others.

Students seeking to become scientists generally envision careers of doing research, not teaching or managing. Graduate and professional schools reinforce this, with most offering little or no instruction on teaching or management. Yet students intending to become professors at least know that teaching will likely be part of their future duties. Management, however, might not loom large as a possibility in the minds of most students, even though it has become a prominent role for many scientists.

Scientists employed in industrial, governmental, and other large organizations and those who primarily conduct grant- and contract-funded research for a living may be particularly destined for research management. In most large organizations that perform research, management is the main path for career advancement. Doing research itself in such settings is lower status, often involving limited authority and salary. For scientists earning their income through research grants and contracts, the pressure to maintain continuous and sufficient funding typically forces many to oversee multiple concurrent projects that are performed by others. In both cases, the opportunity to do research—actively participate in the key activities—recedes or vanishes entirely. Young scientists in particular may struggle with their desire to be actively involved with research as it conflicts with the practical need to restrict themselves to supervision.

Effective research managers can be much more productive, in terms of projects completed and papers written, than scientists who do their own research. This productivity, though, often comes at a price. Managers are removed from the actual research, and thus their knowledge of the status of their projects and the ideas, methods, results, and interpretations associated with them can be less than that of their staff. This distance from the work can lessen managers' ability to have the work done in the way they want, leading to errors in research and a diffusion of accountability. As their own research experience fades into the past, managers may forget the challenges and intricacies of doing research, and develop unrealistic expectations for their projects and staff. Managers juggling supervision of multiple projects may give too little and too infrequent direction to their research staff, impeding progress. Moreover, managers disconnected from the research they lead are at greater risk for letting their preconceptions, ideologies, or other agendas dominate as subordinates defer to their bosses, leading to neglect or distortion of the evidence and good scientific practice.

Managers must also suffer endless meetings and wrangle with bureaucracy. The management role may prevent scientists from enjoying the fun and creativity of actually doing research —the very experiences that led them to become scientists in the first place. Managing research may also lead to atrophy of research skills, as there may be no time for managers to maintain them and acquire new ones. Despite these potential drawbacks of managing, research managers can greatly advance their scientific causes by garnering the necessary resources for research and communicating research to a broader audience.

39

Handling Authorship

... the convention in our lab was that one did not put one's name on a paper unless one had made a significant contribution to it.
—*Francis Crick, 1988*

Authors on scientific publications and presentations get credit and are responsible for the ideas, evidence, and work they report. Probably most scientific research now involves the efforts of multiple persons. Researchers who work together on a project sometimes get into conflict with each other about who deserves authorship or the order of authors for a particular paper or presentation.

The minimum criteria for authorship vary across fields, institutions, individuals, and time. Moreover, the value of a person's contribution to a project is subjective. Researchers who are not the project leader must accept their lesser power if they decide to participate in a project. This leaves them mostly at the mercy of the leader or other researchers with more power when it comes to authorship and other matters. There is no escape from this natural dynamic—it simply is the price of working with others. By choosing carefully with whom you work (see Chapters 35 and 36), you can avoid many problems related to authorship.

Setting clear expectations is essential for handling authorship smoothly. If you are not the leader of the project and co-authorship is important to you, ask whether your work will result in co-authorship *before* you begin that work. If you are the project leader and ask someone else to do something for the project you consider worthy of co-authorship, invite that person to be a co-author then. Among longtime colleagues or friends with high levels of trust, such discussions may be unnecessary, as they may share opinions on authorship or settle any differences quickly.

Project leaders have a special responsibility to include all persons deserving authorship. For instance, it easy to overlook persons who contributed early in a project only, either due to a specific role or because they left the team for other reasons. They still merit authorship, regardless of current circumstances or relations.

Authors without authorship

Lists of authors on scientific publications and presentations are sometimes incomplete. In some settings, contracted researchers and writers effectively do all of the critical work and others (often high status researchers or professionals) attach their names to the work. The contractors usually are listed as junior authors or not included as authors at all. Sometimes the listed authors do not even recognize the contractors in the Acknowledgments (see Chapter 24). Thus, contractors are often paid not only to do the work, but to give up credit for doing it and allow others to take the credit. Another scenario of voluntarily missing authors occurs when researchers decline authorship, even if they did much, or even most, of the work reported or made large intellectual contributions to it. This may occur if a researcher disagrees with his or her co-authors about some aspect of the publication and they cannot find a suitable compromise.

Student research

Mentoring a student while he or she conducts research does not automatically warrant authorship for the mentor on the student's publications and presentations resulting from that research. Mentoring involves suggesting research questions, giving advice and guidance, teaching methods, and editing and providing feedback on manuscript drafts. These are typical responsibilities for professors and other mentors, even in the absence of publications and presentations.

Mentors may claim co-authorship legitimately in specific circumstances, such as when mentors do nearly all the work converting a student's thesis into a journal manuscript, develop new methods or theories for the student's research, or perform key parts of the data collection, management, or analysis (beyond instruction). In these cases, the research represents a sort of collaboration. Similarly, mentors may also be co-authors when the student's work is part of a team project. Segments of a student's thesis may even be separate publications with multiple co-authors, including research staff and collaborators who were not the student's mentors. Naturally, mentors must get students' explicit consent to be added as co-authors.

Nominal authorship

Sometimes, researchers demand or accept authorship for trivial acts, such as contributing previously collected data, giving brief comments on a near final manuscript, or leading the research group in which a study took place. An even emptier category of authorship applies when scientists merely *agree* with what others have written and add their names to the author list. This is especially common in opinion pieces of all types. Persons who claim authorship of work that they paid others to do (see "Authors without authorship" earlier in this chapter) also are nominal authors.

Scientists may offer co-authorship to other researchers in exchange for access to their previously collected data. Or they may offer co-authorship to research partners (see Chapter 37), such as gatekeepers to a resource needed for the research, to enable a project. Such offers, in my opinion, are tolerable compromises to accomplish research, even though they may dilute the meaning of authorship. Other scientists may offer co-authorship, after completing their research, to high status scientists in hopes of boosting the chance their manuscript will be published in a prestigious journal. I believe this kind of offer is unethical because the researchers did not need the nominal authors to conduct the research.

Authorship represents responsibility for some essential *scientific* part of the research. Ultimately, all authors are accountable for their decisions. Authors in name only diminish their integrity by misrepresenting their roles. Others who question them about the published work soon discover their fraud. Take authorship only when you have earned it on scientific grounds.

Some journals require authors to indicate which research tasks each author performed. These unvalidated descriptions often overstate the contributions of some authors, and almost invariably do not indicate the extent to which an author performed a particular task. Nominal authors still slip past these kinds of journal requirements.

Abuses and injustices

Researchers who feel they have been unjustly excluded from authorship often have few good options to rectify the matter. If you have been excluded as an author and want to fight the exclusion directly, you must have good documentation about your contributions and be prepared for a long and public struggle. Such a fight could involve journals, research institutions, and funding agencies and may harm your reputation, even if you are right.

Another possible response is to beat the authors into print, if you learn about your exclusion before the authors have published. If you were responsible for a distinct method, idea, or another intellectual element, you might be able to report it in your own separate paper and publish it quickly in a preprint archive (see Chapter 32) or similar outlet, before the authors' paper is published. This would allow your contributions (provided they are only your own) to be recorded as such in the literature and establish your priority (and the authors' misdeed in denying you authorship). Taking this action would likely damage your future relations with the authors. To pursue this strategy, you must make sure you have legal right to publish the material. If you were under contract to do the work, or you rely on data that are not public and do not belong to you, you might not have that liberty. Regardless of how you otherwise respond to your exclusion, I recommend not working again with the authors who excluded you.

Author order

Customarily, the first author is the one who did most of the critical work and/or led the research. Readers expect the first author to be able to answer all major questions about the research or ideas described in a publication. In some fields, the last author is the head of the research group in which the research was done, the supervisor of the overall project, and/or the person who secured funding for the research. In such cases, the last author may have done little of the actual work reported and may be a nominal author. Generally, the sequence of other authors indicates their levels of contribution in descending order, although sometimes researchers use other ordering schemes. In practice, differences in authors' power and status may also influence author order, such that powerful authors get more prominent author positions than they might otherwise deserve. If you are frustrated with repeatedly receiving secondary positions in author orders, seek to publish some work on your own or lead research in which you control the authorship decisions.

40

Contacting Other Scientists

When I walk along with two others, they may serve me as my teachers.
—*Confucius, 480–350 B.C.*

If with pleasure you are viewing any work a man is doing,
If you like him or you love him, tell him now ...
Do not wait till life is over and he's underneath the clover;
For he cannot read his tombstone when he's dead.
—*Berton Braley, 1915*

The opportunity now to learn from other scientists is better than ever in history. Libraries with extensive collections exist throughout the world, scientists gather regularly at meetings and conferences in every region, and Internet access spans most of the planet, with scientific resources at almost everyone's fingertips. Your scientific horizons will grow greatly if you expand your scientific contacts beyond those researchers whom you already know personally. These connections open new channels for the flow of knowledge, both to and from you.

Contacting scientists whom you don't know can serve several practical purposes. If you are unable to obtain a publication or unpublished report for free online or from your library, request it directly from the author. Most scientists are glad to send their work to other interested researchers. (Books are an exception: buy them or borrow them from libraries or inter-library loan services.) Contact other researchers to ask them questions about their work, if you do not understand something in one of their publications. Another reason to contact other scientists is to request data which they have not archived publicly but that you would like to analyze (see Chapters 9 and 10).

Requests for information are not the only reasons to contact other scientists you don't know—you can also send information.

As researchers, we rarely get much positive feedback. If you come across work that you find especially helpful or insightful, thank the author directly for his or her contribution. It takes little time to do and can certainly boost another scientist's spirits. One of the first times I sent a complimentary message to an author I did not know was when I was a young researcher. I told the author, a senior professor at a somewhat obscure college, that I thought one of his recent articles was particularly interesting and well-done. He wrote back that my note had made his year. If you sincerely feel good about another's work, why not spread some cheer?

Send your publications to other researchers who—whether you know them or not—might be interested in them. Such persons might include, for example, many or all of those you cited in a publication, others who publish on the same or similar topics as your research, and potential critics. When you send your work to others, be sure to ask them for any comments or criticisms they may have about it. This is like a further informal round of peer review in addition to the feedback you seek on your research proposals, protocols, and unpublished manuscripts (see Chapter 31).

I first encountered many of my collaborators by contacting them long-distance, as a stranger, for one or more of the purposes I have discussed here. Some of these collaborators are also now among my closest friends. Although such positive outcomes are an additional reason to contact other scientists, keep your expectations low. Many researchers do not respond despite repeated contact attempts, and those who do respond often may not grant your requests.

Making contact

Science is a global pursuit, and most scientists whose work in which we are most interested do not live near us. For many topics, there are only a handful of active researchers worldwide, and unless you initiate contact, you may never have a chance to communicate with them directly. Compared to most other professions, scientists have fairly informal patterns of interaction, with few restrictions based on status. I began contacting other researchers when I was an undergraduate student and have continued to do so throughout my career. It is legitimate to contact any scientist, regardless of your background or the other scientist's background. I've contacted students, non-professional researchers working outside of academia, high status professors at famous universities, and other scientists working in almost every imaginable context all across the world. You don't need to be extroverted to contact other scientists, but you do need to be interested in your topic and eager to learn from others.

Contacting other scientists you don't already know usually occurs long-distance by email, post, telephone, Internet/video calling, social media, or fax. When you have identified someone you want to contact, usually you will also have the information on how to reach that researcher through one of these modes of communication. However, if you lack such information, try these strategies: do a web search with the person's name and name of his or her organization (if any); do a literature search (see Chapter 5) to find publications that list the person's contact information (if she or he is listed as corresponding author); and search for the person on relevant social media. If you still cannot find any information to contact the person directly, try contacting the researcher through intermediaries whom you might have found in your various searches, such as co-authors or administrators at their organizations.

Before making contact, review the researcher's web page (if available) and publications to get a sense of his or her interests, work, and history. I recommend that first contact attempts be done by email, post, social media, or fax, as they are less intrusive than calling by telephone or Internet. Unless you know otherwise, always address other scientists as Doctor or Professor (the latter if connected to a university or college). Some scientists are very status conscious and would be offended not to have their educational achievements or positions recognized. If the researcher actually is not a professor or does not have a PhD, MD, or similar degree, you have done no harm; he or she may even be flattered by your mistaken assumption. Write your message or letter clearly and concisely (to respect the other scientist's time), and proofread it. I suggest making two or three contact attempts spaced at least one week apart; if the researcher does not respond, then try another mode of communication. In all of your contact attempts, be courteous.

You may also be able to contact a particular scientist at a conference. Ask a colleague who knows you both to introduce you, or find out when the other scientist is presenting, and approach them afterwards.

The best way to learn from other scientists, of course, is in person. Conferences give one kind of opportunity. If you are able to visit other scientists where they work (usually after first contacting them through other means), it is much better. Making or receiving such visits is uncommon, however, unless you or your visitors are well-funded.

Another way to contact other scientists generally, if not particular individuals, is by posting to forums, comments sections in blogs, email lists, or other online groups. Such online communities can play a vital role for connecting researchers, and sometimes interactions in these settings develop into direct communication between individual scientists. However, participants in such online groups may not necessarily be representative of scientists in your field or include those researchers most beneficial for you to know.

If you want to learn from other scientists, especially during your first interactions with them, focus on asking questions and understanding their responses. If you dominate the conversation, not only do you learn less, but you signal a lack of interest in the person with whom you are communicating. Listening respectfully doesn't necessarily mean agreeing—you can still challenge politely. The key is to engage positively and productively with your new colleague.

41

Responding to Other Scientists

Manners must adorn knowledge, and smooth its way through the world.
—*Philip Stanhope, 1748*

... we gave to those interested whatever information they asked of us.
—*Marie Curie, 1923*

The complement of initiating contact with other researchers (see Chapter 40) is responding to other researchers who contact you. Both are essential to the free flow of scientific knowledge. By publishing (making public) your scientific work, you become a public figure of sorts. Almost every scientific publication includes the author's contact information (or that for at least one author on multi-authored publications). Your entry to the public arena entails obligations to your fellow scientists. We are in the business of creating and sharing information, and it is our duty to make that information accessible.

The benefits of responding to other researchers are the mirror images of the benefits of initiating contact: spreading knowledge of your work; clarifying your published work; allowing others to confirm and extend your work (through sharing data); learning about others' research; and potentially gaining a new colleague or collaborator. If you do not respond to communication from other scientists, you deprive yourself of these benefits and may also damage your reputation from not meeting your responsibilities. Researchers who contact you are likely to be doing work related to your own. Not only could they be potential collaborators or someone from whom you could learn, they could also be potential students, reviewers, or editors. Whether and how you respond is effectively public behavior, particularly with respect to requests for data or information about your published research. Your response (or lack thereof) may thus become public knowledge, documented in a future publication by a requester. Thus, not responding or responding poorly can come back to hurt you in the future.

In my career, I have contacted many thousands of scientists in diverse fields. I estimate that I have received responses from only about half of them, no matter whether my contact was a compliment, question about their work, request for a publication, or request for data.

Show simple courtesy and respond to all genuine communications you receive from other scientists. Be grateful for any interest in your research. Keep copies of all of your past publications, presentations, and other scientific documents readily available so that you can send them to requesters. To reduce the number of requests and ensure that none are forgotten, make all of your future publications available to everyone by publishing on preprint servers (see Chapter 32) or in journals that make their content available to all for free online. Similarly, when you finish current and future projects, make your data available in public archives (see Chapter 9).

42

Giving Comments

… the peculiar evil of silencing the expression of an opinion is, that it is robbing the human race, posterity as well as the existing generation; those who dissent from the opinion, still more than those who hold it. If the opinion is right, they are deprived of the opportunity of exchanging error for truth; if wrong, they lose, what is almost as great a benefit, the clearer perception and livelier impression of truth, produced by its collision with error.
—*John Stuart Mill, 1859*

Science is the search for truth—it is not a game in which one tries to beat his opponent, to do harm to others.
 —*Linus Pauling, 1958*

The point of reviewing is to ameliorate.
—*John Potterat, early 2000s*

The greatest gift one scientist can give to another is a thoughtful critique of his or her ideas and research. In fact, we *owe* fellow scientists our critical opinions, especially when they request them and often even when they do not. Science advances primarily through criticism.

Giving comments also benefits the commenter. Reviewing others' work is good practice for assessing methods, interpretations, writing styles, and other aspects of research reports. You can then apply the same critical lens to your own work (see Chapter 61). Reviewing may expose you to new ideas and types of research, too.

We give comments to other researchers on manuscripts, study plans, project proposals, and study protocols, among other kinds of documents and materials. Sometimes those requesting comments are the researchers themselves; at other times, the requester is a funding agency, the editor of a journal, or someone evaluating the researchers involved. Much scientific correspondence and written debate (see Chapter 43) also involve giving comments.

Courtesy

There are several golden rules of reviewing. One is to review others' work at least as much as you ask others to review your own. If you agree to review, do so within the time frame given by the author or person requesting the comments. This also increases the chance that the author or requester will reciprocate in the future.

Structure and substance

I recommend starting comments with a brief summary of your feedback. Highlight the main strengths and problems as you see them. No matter how bad you think the work might be, note at least one positive aspect before mentioning the problems.

Organize your specific comments by matter and manner. *Matter* refers to the scientific content. *Manner* refers to how the author(s) presented the content, including writing style, grammar, spelling, punctuation, and formatting. While both matter and manner are essential, discuss matter first, because it is more important.

In your comments, distinguish objective information from opinion, whether your own, the author's, or others'. Criticize misrepresentations, illogical ideas, or statements that are clearly inconsistent with empirical evidence, but otherwise support an author's freedom to opine and speculate in the appropriate sections of a report. Similarly, make your opinions and preferences explicit with accurate modifiers, such as "in my opinion" or "I recommend" (see Chapter 27).

Other useful comments include references to resources that will help the author fix problems you identify, if only in future research. Also, for each specific comment, identify the location (at least page number) in the document to which it pertains.

Another golden rule of reviewing is to hold your comments to the same standard you expect an author to meet in his or her writing. Therefore, proofread your comments for spelling, punctuation, grammar, coherence, and clarity.

Be careful not to change the status of the author's research with your suggestions. For instance, reviewers of research reports often call for revised or further analyses of the author's data. If the author planned his or her analyses to evaluate an hypothesis, such additional analyses may amount to exploration, if they are informed by the results themselves (see Chapter 51). (However, requests for more complete descriptive analyses in descriptive research and analyses focused on substantively different questions not posed by the author do not affect the status of the author's work.) Once researchers report their analyses, it may be too late to change course and maintain integrity of the original research plan. Thus, traditional peer review for scientific journals occurs at a point when comments may be irrelevant for the *conduct* of the research reported in a manuscript. Otherwise, comments can be useful guidance for *future* studies or aspects of the report that can be changed constructively, such as discussion of prior research, interpretations, implications, and limitations. Given these constraints, it is especially important to seek and give comments on study ideas, plans, and protocols to be pre-registered (see Chapters 4 and 31).

Tone

A third golden rule for reviewing is to give comments in a way that you would want to receive them if the document you are reviewing were yours. Feedback and advice are helpful; demands and insults are not. One way to fulfill this golden rule is to qualify your comments by stating them tentatively ("it seems", "I think", "may", "might", etc.). As readers, we cannot be certain that we understand the author completely or know his or her intent. Phrasing some comments as questions emphasizes this uncertainty and also frames points for the author to address directly. As reviewers, we can feel like judges of authors, making it easy to forget that we ourselves are fallible and some of our comments might be wrong. Reviewing anonymously especially facilitates such feelings of superiority. One way to counteract this in such circumstances is to identify yourself in your comments, as in signing them.

Censorship

Communication between scientists who are free to decide how to conduct and report their own research allows high quality science in the long run. A scientific reviewer is a critic, but not a censor. Just as it is imperative we publish all of our work (see Chapter 32), it is equally crucial not to deny others the same access. There are no defensible scientific or ethical reasons to prevent others from publishing, even from publishing work that is seemingly nonsensical or riddled with error. One way to address such work is with publishing other research or commentary in response (see Chapter 43). None of us can *know* the ultimate value of another's work, either now or in the future, nor is it anyone's responsibility or right to decide who may be allowed to communicate their research or ideas. The current dominant mode of scientific publishing in which editors of journals accept or reject submissions for publication, usually based on the recommendations of reviewers, rests on the belief that such judgments should be made.

Ideas and evidence cannot be dangerous or harmful in science, or perhaps in any context. There can be good ideas and bad ideas, and strong evidence and weak evidence. Some ideas and evidence may, however, be threatening to those defending other (possibly worse) ideas and (possibly weaker) evidence. But these aren't dangers to science. On the contrary, they are essential to science.

43

Engaging in Scientific Debates

Be calm in arguing: for fiercenesse makes
Errour a fault, and truth discourtesie.
—George Herbert, 1633

Since the method of science is that of critical discussion, it is of great importance that the theories criticized should be tenaciously defended. For only in this way can we learn their real power. And only if criticism meets resistance can we learn the full force of a critical argument.
—Karl Popper, 1963

But we must not expect too much; we must not expect that a confrontation, or even a prolonged discussion, will end with the participants reaching agreement ... victory in debate is nothing, while even the slightest clarification of one's problem—even the smallest contribution made towards a clearer understanding of one's own position or that of one's opponent—is a great success.
—Karl Popper, 1976

Scientists often exchange divergent views with each other on some topic in explicit, focused discussions. These debates may or may not be part of broader, long-running disputes in a field, such as between rival schools of thought, theories, hypotheses, methods, or approaches. Debates can occur in person, for example, in the question and answer period after a presentation, formal debates at a conference, or informal conversation with other researchers. Debates in print typically involve three stages: an original contribution, often a research report; a letter to the editor or online comment by a critic of that contribution; and a response from the original author(s). Many of the elements of debates are also present in other interactions between scientists, such as critiquing prior work in publications and presentations, giving comments on manuscripts (see Chapter 42), and responding to journal reviewers' comments. Most circumstances of debates involve, nominally, a critic commenting on another researcher's work. However, criticism in such exchanges is usually mutual, with both sides criticizing each other's ideas and evidence.

Purposes of debates

No side wins a debate. Participants in scientific debates rarely change their minds as a result of their opponents' arguments. Debates also seldom shift non-participating researchers' opinions, which may be determined more by prevailing beliefs, funding priorities, and politics. Seeking victory in a scientific debate often leads to disappointment or delusion.

If persuasion is not a realistic goal, why participate in a debate? For critics, challenging other scientists publicly may be the only way to get them to respond to questions and requests if they are not responding privately (see Chapter 41). In other cases, a researcher may make a claim that a critic thinks is inaccurate, and the critic may seek to correct or clarify the matter in public. Or a critic may wish to offer an alternate hypothesis or explanation for a result reported by another scientist.

Publishing or presenting your work is a public invitation to debate. Responding to critics is part of doing research. If you do not respond, others will come to their own conclusions. As Sophocles (c. 497–405 B.C.) observed, "Know'st thou not, that silence but admits the accuser's charge?" Even if your response is to agree with your critic, replying makes clear to others what your views are.

The most important reason to participate in a debate is to discover the strengths and weaknesses of ideas and evidence, both your own and others'. Scientific scrutiny is a scarce resource, and every bit our work receives is to our benefit. Through debate we can also learn to describe better the views of all involved. No matter the purpose, all debates make the scientific record more complete and thus may benefit future scientists in ways difficult for current scientists to predict.

Although we normally should attack our own ideas and evidence (see Chapter 61), our role in a debate is different. In this context, we must explain our views, highlight their value, and defend them against misrepresentation and dismissal. Similarly, guided by reason, we must attack weaknesses in our adversaries' ideas and evidence. As Popper emphasized, both vigorous criticism and defense are necessary to pursue truth.

Harmful tactics

Behavior in debate reflects character. Preserve your integrity and the benefits of debate by avoiding tactics that discredit you and your views in the eyes of rational observers. Many of these harmful tactics are informal logical fallacies (see Chapter 56).

Sometimes researchers claim the debate should not take place, or at least not publicly, because it could it could damage public opinion of the field or profession, lead people to think or act in ways the researchers don't like, or make funders reluctant to sponsor research on the topic. This ploy is plainly anti-scientific, as open communication, including debate, is fundamental to science.

Never infer an adversary's motivations. These are not possible for you to know with certainty and are irrelevant in debate. Only the *quality* of ideas and evidence matters, not what inspired them. Similarly, the source of a researcher's funding, or lack thereof, is not pertinent. If anything, researchers handicap themselves in debate and the pursuit of truth by having a sponsor or employer to please (see Chapters 16 and 62).

Do not disparage other researchers' lack of prestige, training, or even experience, as they are also separate from the content of their ideas and evidence. Even non-scientists sometimes contribute usefully to scientific debates. Likewise, touting your own status, education, or experience (apart from giving relevant evidence) is really an argument from authority and has no place in a scientific debate. In addition, praising or denigrating research based on where it was published or presented is immaterial in debate. No context for communicating scientific research validates or invalidates the work (see Chapter 32).

Avoid other kinds of ad hominem attacks (attacks on the person, not the argument) and name-calling. Currently popular scornful labels include, among others, "pseudoscientist," "crank," and "crackpot". "Denialist" or "denier" is a particularly offensive insult, likening dissenters and skeptics of a particular view to those who deny the Holocaust happened. The real deniers of science are those who call others deniers, thereby suppressing debate. Referring to another's work as, for instance, pseudoscience or denialism is also an ad hominem attack, just phrased indirectly. All of these labels are subjective. Why would a scientist trying to pursue truth ever use such derogatory terms that divert from rational debate of ideas and evidence?

Shun veiled ad hominem attacks. For example, some researchers try to smear their opponents for their unpopular or unusual views on topics beyond the scope of the debate, attempting to stain, by association, opponents' views that are related to the debate.

Sometimes participants in a debate tell opponents that they are "entitled to their own opinions, but not their own facts." This sounds like an accusation that the opponents are making up facts. However, often what the relevant facts are may be in dispute. This accusation is actually contradictory: scientists can have different opinions about what the facts are and how they should be interpreted and weighted.

Critics are not obligated to present solutions to problems that they identify in others' work. Noting problems and proposing alternate solutions are logically separate. Dismissing critics because they do not also present replacement solutions is therefore an irrational tactic to invalidate both the critics and their criticism.

It is inappropriate to brand adversaries' work as "irresponsible" or "dangerous" or otherwise condemn their research. These are the sentiments and words of political repression and do not belong in any scientific debate. Calls for journals to withdraw objectionable (in the eyes of some) papers or censure editors for publishing them are anti-scientific and poison the pursuit of truth, regardless of the papers' content or quality. Scientists have an ethical *duty* to publish their research (see Chapter 32). Blocking scientific publication of any kind, however, is unethical and inhibits the development of reliable scientific knowledge.

Similarly, avoid scaremongering and constructing fantasy worst case scenarios of potential impacts of another's research or ideas. Nobody knows the future, or how *anyone's* research and ideas will affect it. Sounding the alarm about possible threats that others' research poses is disguised coercion, a rallying cry for journals, funders, governments, and others to squelch and halt another's scientific work.

Mean-spirited remarks of all kinds in debate often have a chilling effect. Extremely hostile attacks can discourage researchers, especially inexperienced ones, so much that they quit science or avoid conferences and meetings, particularly those in which prior critics attend. When debate becomes so bitter that it stops, the pursuit for truth also stalls, as Herbert recognized.

Helpful tactics

You can take several steps to make your side of the debate as productive as possible. Aim to represent others' views accurately. If your debate counterparts claim that you have misrepresented their ideas or work, apologize and correct your course. Play devil's advocate to sharpen your understanding of another's views. When preparing your criticism of another's views, imagine how he or she would respond to your critique. When preparing a description of your own views, imagine how a critic would attack them. If you are unsure of how adversaries might think about some matter, read what they have written on the topic more thoroughly. Be able to explain and defend your opponent's views well, so that you can preempt many of their criticisms.

Focus on the content of the debate, and not your counterpart's debate tactics. If your opponent argues erratically or uses in unprofessional tactics (see "Harmful tactics" above), don't mention these failings apart from making your and their arguments clear. It can be difficult to ignore insults, dismissals, and irrational arguments, but it is essential to stay focused on ideas, evidence, and reason. Don't let your opponent's emotion and illogic lure you into destroying your own arguments with the same. You can further objectify the debate by giving descriptive names to ideas, hypotheses, and methods, rather than referring to them as "my", "our", "your", or by the names of the researchers who proposed them (see Chapter 42). This helps concentrate the discussion on ideas, not persons.

In responding to another's criticism, address each point directly, fully, and fairly. Lack of response to a point of criticism is equivalent to concession and also signals an evasive attitude. If limits on the length of your response in print or online prevent your addressing each point, then give your full replies in online supplementary material, a personal web site, or some other published source. Whenever possible, address points with sound empirical evidence, and acknowledge gaps in evidence freely.

Scrutinize your own views for common flaws in reasoning (see Chapter 56). If your adversary makes a good point or identifies an error of yours, thank him or her for it. Similarly, when your adversary clarifies a point and thereby eliminates your dispute about it, affirm the resolution.

Be charitable when debating with a presenter. Not everyone can think well under pressure and in front of an audience. It is easier to ask a question or give a comment from the audience than it is to respond in front of an audience. Unintentional misstatements and omissions are common during unscripted oral debates. Written statements, however, reflect a scientist's careful, considered views and are fair targets for criticism. Nonetheless, spontaneous spoken comments can be quite revealing, especially with respect to a researcher's potential biases.

Charges of scientific misconduct are legitimate matters of scientific debate. Accusers should be free to accuse (with solid evidence, if only to protect their own reputations) and the accused should be free to defend. However, accusations of misconduct can quickly lead to acrimony and unproductive debate. Most topics that relate to misconduct can be debated in a more effective, de-personalized way by focusing on anomalies and deviations from good scientific practice.

Ultimately, scientific debates are sterile unless scientists conduct new research to address disputed matters. Therefore, propose specific empirical tests that can reveal the relative merits of different views: types of new studies to conduct, additional analyses of existing data to perform, further relevant information for authors to report on their published research, and/or data that underlie the debate to archive publicly. Regardless of which side of a debate you are on, if you continue to do research on the topic, show your integrity by performing these empirical tests rigorously. Unfortunately, though, defenders of the status quo usually ignore these challenges or do so insincerely, thus critics often are the ones who ultimately do this work.

Proponents of established, dominant views tend to forestall debate on them. If you are a critic, persist in getting your views published. If a journal editor rejects your letter to the editor or online comment, publish your commentary elsewhere, such as a preprint server or post-publication peer review site.

44

Chairing a Conference Session

And let him be sure to leave other men their turns to speak. Nay, if there be any that would reign, and take up all the time, let him find means to take them off, and to bring others on ....
— *Francis Bacon, 1597*

When you chair a conference session, you are director of the show, even if you did not organize the session. As a chair, you are a representative of the conference. You must take control: both presenters and the audience look to you to lead.

Preparation

Try to meet with presenters individually before the session starts. This is especially helpful if you do not already know them or even know what they look like. When you talk with them, ask whether they have any specific needs or requests of you and how to pronounce their names (if you are uncertain). Tell them about the organization of the session—how much time they have in total, when they can take questions, and your method for alerting them as to how much time they have left. For these alerts, I recommend making written or printed signs that you show to presenters silently.

If the conference provides audio-visual, computer, or other equipment for presenters, ensure that they are working and identify conference staff who can help with any technical problems. Check that the physical arrangement of the room, including seating, lighting, and temperature are suitable, and make adjustments as necessary. Also, if the conference provides water in the conference rooms, make sure some is near to presenters while they speak.

Helping presenters and the audience

Don't make introductory remarks at the beginning of the session or concluding ones at the end (aside from something like "Welcome to the session on ...." and "Thank you for attending this session") unless the conference schedule has slots for them. Chairs who make such unscheduled comments simply steal time from the presenters. Introduce each presenter by name and give the title of his or her talk. During a presentation, be prepared to assist the presenter in any way, such as fixing equipment that malfunctions (or getting technical help immediately), advancing slides, or passing out handouts. Similarly, monitor the environment of the room and modify the lighting, temperature, and noise (by shutting doors or windows) as needed.

Regulating time

Ensuring that each presenter has the scheduled amount of time is a matter of fairness not only to presenters but also members of the audience who move between sessions expecting to hear particular presentations at their scheduled times (for those conferences with individually scheduled presentations). Be a strict timekeeper and keep to the schedule. If for some reason a presenter cancels a presentation, wait to begin the next presentation until the scheduled time. And if one presenter starts later than scheduled for reasons outside of his or her responsibility, let the speaker use the full *duration* of the time slot (meaning that the following presentations may also start late). Give each presenter his or her allotted time and no more, regardless of who the presenter is—even if he or she is a Nobel laureate or your friend, colleague, boss, or student. Give presenters a visual signal of how much time they have left at pre-determined time points (such as with 10, 5, and 1 minutes remaining). Let presenters choose how to divide the remaining time between their presentations and question and answer periods. When the presentation time is up, interrupt if necessary, thank the presenter, and announce the next speaker. If you are presenting in the session, find someone to keep time for you and alert you to the time remaining. Respect the time limit for your presentation—you don't get extra time or flexibility because you are chair—otherwise, your credibility and authority as chair vanishes and your fellow presenters may resent you, justifiably.

Question and answer period

If you are able to set the rules for the session, insist on having questions and comments from the audience after *each* presentation. Postponing the question and answer period for all presenters until after the *last* presentation in a session shortchanges the earlier presenters, who rarely receive questions and comments in that format. Let audience members (who may leave after the presentation) and presenters interact while the material is fresh in their minds. Allow presenters to select questioners, but you can call their attention to questioners they might have overlooked. End the question and answer period when the speaker's allotted time is up. If a presenter has used up his or her whole presentation time before receiving questions, just announce the next speaker after explaining the presenter didn't leave enough time for questions.

Critical discussion is perhaps the defining feature of science, as Karl Popper emphasized repeatedly in his work. A key job for session chairs is to facilitate such discussion, no matter how heated it gets. Sometimes, argument erupts, tempers flare, and accusations—even insults—fly. Nonetheless, chairs are not responsible for the behavior of presenters or the audience. If intense discussion is still going when the presenter's time is up, invite all involved to continue it with each other after the session.

For each presenter, always have one question, no matter how small, that you can ask, in case no one else has one. Questions are a sign of interest from the audience. Presenters spend much time and effort to prepare and give their presentations, and they deserve respect for that. Try to prepare the questions before the session, using presenters' abstracts as inspiration for ideas.

Intellectual and Ethical Matters

45

Chasing Wild Geese

Ideas develop spontaneously in the mind, and when one yields to his thoughts, he is like a man at the window watching the passersby. In some such fashion one watches his ideas pass by. This requires no effort, and it even has great charm. Where the work is, and the fatigue, is to collar the idea, like one stops the passerby despite his desire to flee, to retain it, to fix it and give it its character.
—Claude Bernard, 1850–1860

If you want to have good ideas you must have many ideas. Most of them will be wrong, and what you have to learn is which ones to throw away.
—Linus Pauling, 1901–1994

All great scientists have used their imaginations freely, and let it ride them to outrageous conclusions without crying 'Halt!'
—Jacob Bronowski, 1967

Thinking about a scientific matter in an open-ended, unstructured way for an extended period is what I call a wild goose chase. The path your thoughts take in a chase is unpredictable and may change suddenly. At the outset, the final destination of the chase is unknown. This kind of wild goose chase is an essential part of scientific work, as good research involves as much *thinking* as *doing*.

Wild goose chases can focus on solving problems of diverse types; interpreting results; developing hypotheses; creating methods, designs, or analytic approaches; conceiving and planning studies; and composing text for reports or other documents. Another worthwhile kind of chase is conducting a thought experiment, imagining various outcomes of potential or planned research (see Chapter 4), their implications, and what research might follow them. Sometimes the excitement of developing a project obscures a realistic appraisal of the project's scientific value—how much does it truly advance scientific understanding?—and a wild goose chase can be a way to assess this. A further valuable kind of chase is trying to destroy your ideas and interpretations, and develop alternatives (see Chapter 61).

Just like reading the literature (see Chapter 47), going on wild goose chases helps keep you intellectually fit, regardless of outcome. In their autobiographies, many scientists describe their own wild goose chases.

When you go on a chase, let your imagination run wild, associate freely from one idea to another, pursue tangents, and censor nothing. When you come to sticking points or problems, postpone your internal criticisms until you have developed your ideas more fully. After that is done, then you can scrutinize any questionable ideas that you developed earlier. During a chase, take notes on each idea that seems firm at the moment (see Chapter 46), so you don't forget it. These notes also serve as reference points as the chase continues.

Many wild goose chases, as the name implies, do not yield tangible advances. Nonetheless, the advances will come as you go on more chases, in line with Pauling's advice. Some wild goose chases can be exciting. You might feel like you are hot on the trail of a major breakthrough or insight, no matter what the prospects of success may be. I have occasionally gone on a chase for hours at a time, with nothing more to show for it at the end than a few notes and maybe a sense of progress (if I feel like I resolved something) or unease (if a problem remains). Any remaining unease is actually good, as it will fuel a chase at another time to push toward some sort of resolution.

Wild goose chases can be done together with one or more colleagues. This can be especially fun, like playing mental pinball with a partner. To facilitate their thoughts on a chase, some researchers go for walks, while others look out a window or play a simple game such as bouncing a ball. I try to have at least one wild goose chase ready to go at any time. There never is a reason to be bored or want for entertainment—go on a chase for a bit while waiting somewhere or doing some routine activity that doesn't require your full attention. Some wild goose chases may recur, taking days, weeks, months, or years to run their courses.

Be careful about when and where you go on chases. Chasing wild geese takes all or most of your attention, turning you into an absent-minded professor, lost in thought. So don't go on a chase when you are driving a vehicle or in a circumstance which you must focus on your surroundings for your and others' safety. Similarly, don't go on a chase in a social setting requiring you to be present mentally, as non-scientists might think you are doing nothing and expect you to interact with them normally. Likewise, avoid interruptions of all kinds for a productive chase.

46

Making Notes

I had ... during many years followed a golden rule, namely, that whenever a published fact, a new observation or thought came across me, which was opposed to my general results, to make a memorandum of it without fail and at once; for I had found by experience that such facts and thoughts were far more apt to escape from memory than favorable ones.
—*Charles Darwin, 1809–1882*

No more great ideas down the drain.
—*Aqua Notes motto*

For scientists, the written word is indispensable. It allows us to record many details that would otherwise be impossible to remember. The written word also is a tool to capture an elusive prey—the fleeting thoughts and insights that are crucial in research. Human memory is fragile, at any age, and is especially so for researchers engaged in many different activities. Thus, making notes is an essential daily scientific task.

Make notes on any aspect of your scientific work, from the mundane and logistical to the intellectual. The notes can serve as reminders to do something or as kernels of ideas to develop further. As Darwin recognized, the most critical ideas to note are those which are least likely to recur: thoughts and observations that run counter to your expectations and views.

There are several circumstances in which you can always anticipate making notes, such as mental "wild goose chases" (see Chapter 45), field observations (see Chapter 2), and research-related meetings. Keep a way of making notes with you at all times, literally, just as Leonardo da Vinci did. Notes can be made in writing (on paper or with an electronic device) or by voice (with some sort of recording or speech-to-text conversion device). Make sure you have a way to make notes at your bedside, as ideas often come just before going to sleep and after waking up. A pad of waterproof paper and a pencil are perfect for making notes in the bath or shower.

For the notes to be useful, review and act on them regularly, such as in updating a to-do list, adding a note to a file, pursuing a topic on a wild goose chase, or discarding an idea entirely. A lost or neglected note is a lost idea.

I start files with notes for projects many months or years before I actually begin to plan and design them formally. I do the same for papers I intend to write based on studies I am currently doing. Some of these projects and papers never materialize, but that is not important. The process of making and reviewing the notes helps me eliminate weak ideas. And when the projects and papers are worth pursuing formally, the accumulated notes mean that much of the writing and intellectual work is already done (see Chapter 18).

47

Reading the Literature

Miss not the discourse of the elders.
—*Ecclesiaticus, 8:9*

Accurate reading on a wide range of subjects makes the scholar.
—*John of Salisbury, 1159*

Scarcely any specialist of to-day is really master of all the work which has been done in his own comparatively small field.
—*Karl Pearson, 1892*

So many people today—even professional scientists—seem to me like somebody who has seen thousands of trees but has never seen a forest. A knowledge of the historic and philosophical background gives that kind of independence from prejudices of his generation from which most scientists are suffering.
—*Albert Einstein, 1944*

Scientists often contend they know the literature on some topic or field. This is a delusion. If such mastery was impossible in Pearson's day, surely it is now as well.

Yet even if complete knowledge of the literature is elusive, there is still great benefit from reading it, with different strategies and practices required for different purposes. By "reading the literature," I mean reading scientific works for reasons other than just preparing a particular report or document of your own.

Reading the literature is best done on a regular basis. By reading the literature, you can learn the history and context of a topic on which you are working that you ordinarily would not address in your own papers. A reading habit also lets you stay abreast of current topics and trends in your field and permits you to assess the evidence on topics in which you are interested but not (yet) studying in your own research. Furthermore, reading the literature is an excellent way to broaden your horizons, stimulate your curiosity, make connections that are missing in your own and others' work, and get new ideas for research.

The most important reason for reading the literature may be for scientific and intellectual exercise. Whenever reading the literature, practice your reasoning skills: generate alternative interpretations and explanations of results, and identify problems with the design and methods and imagine solutions. Reports of observational research, in particular, are ripe for these sorts of critiques. As you read, you can observe logical flaws, poor writing, missing arguments and information, and other defects, which serve as reminders of mistakes for you to avoid. Reading the literature also allows you to learn from good examples. You can discover other ways of framing and solving scientific problems, new methods and designs, creative applications of logic, productive resolutions of scientific controversies, and clear and effective writing styles. In short, reading the literature helps keep you intellectually fit.

To get the greatest benefit, read widely. International, cross-disciplinary magazines and journals such as Science, Nature, Science News, Scientific American, and American Scientist offer exposure to work in many different fields. Online science news aggregators such as ScienceDaily and Real Clear Science do the same. Focus on articles and news stories written for non-specialists or non-scientists, but also sample articles written for specialists to get a glimpse of actual reports in other fields, even if most of the material is incomprehensible to you. The research featured in these sources may not be representative of the corresponding fields—usually, there is a bias toward mainstream and/or sensational work that resonates with readers' fears, hopes, and fantasies. Moreover, journalistic coverage may not be accurate, balanced, or thorough (see Chapter 34). And these publications often have political agendas or cater to those of their readers. Regardless, the articles in these sources still provide opportunities for scientific and intellectual exercise, and they are convenient entry points into the literature.

Some scientists may question the value of reading widely, given that they have limited time, and that their efforts might be best focused on reading in their narrow specialties. Reading the literature is something you do *beyond* what is strictly necessary for conducting your immediate research projects. For intellectual exercise, traveling over familiar ground has little value, and may actually reinforce some questionable views and practices maintained in a specialty.

Browsing journals and preprint archives is a good way to explore particular fields. Examining titles and abstracts of articles sometimes may be sufficient for monitoring trends and topics if your prior knowledge of the field is deep enough. Also, read books, including popular science books (especially those written by scientists). Books can give helpful introductions to a field. Book authors often write in a more informal and personal way and may cover topics more comprehensively than they do in conventional research articles. Researchers' blogs are another source of similar views and information. When you identify a topic that you want to read about more extensively, search the literature systematically (see Chapter 5). Forward and backward citation searches can be particularly useful for finding work related to an interesting source.

When you seek to understand the evidence on a specific topic, consult meta-analyses and systematic reviews (see Chapter 15). You can even conduct your own, perhaps informally, if none exist or existing ones have been done poorly. Any other approach will result in an incomplete or biased view of the evidence. Similarly, single studies are not definitive evidence on any matter, and thus you are better off not trying to remember their particular results. Use reports of single studies for any of the other reasons for reading the literature.

Decide whether to read a specific source based on its content only, rather than the prestige of the author or outlet. On average, high status journals publish less reliable and less rigorous research than medium and low status journals.

Likewise, scientific research is not monolithic on any topic. As you read the literature on a subject, seek minority views and dissenting authors to get a sense of the uncertainties.

When you read individual empirical reports, skip the introduction and discussion sections. Authors sometimes oversell their work in these sections, misleading readers about the methods, results, interpretations, and implications of their research. Instead, focus on the methods and results sections, and come to your own conclusions first before considering the authors' view. Sometimes the methods section alone contradicts claims the authors make in the abstract or title, given that study design and procedures determine what can possibly be known (see Chapter 3).

Be sure to read old research. Old work shows how a field developed and can indicate how much (or little) it has changed in terms of prevailing assumptions and presumed knowledge. It can be exciting to read reports by and thoughts of scientists from many years, generations, or even centuries ago. It is a sort of time travel to learn from our forerunners and fellow seekers of truth. Even though their methods and theories may be unsophisticated compared to the current time, earlier scientists may have identified key problems that remain today. Often, they made critical observations and had powerful insights that scientists in the time since have ignored due to hubris (see Chapter 61) or corruption (see Chapter 62). Reading old work enables us to uncover past mass violations of good scientific practice and rectify them.

Some knowledge of statistics and mathematics lets you understand key parts of research reports in many fields even if the substantive and other methodological content is unfamiliar to you. Keep in mind that the same or similar statistical and mathematical methods sometimes go by different names in different fields.

We scientists are participating in a long-term, inter-generational conversation with other scientists. To contribute to this conversation most effectively, read at least as much as you write. As in any conversation, listening is just as important as speaking.

48

Choosing Productive Scientific Problems and Questions

Research starts with selecting a general *problem* to study. Problems usually are more specific than the broad topic that defines a field or sub-field, and sometimes researchers in different fields study the same problem. In an individual study, scientists investigate one or more research *questions* related to the problem. Questions usually correspond to the goals of a study and sometimes the particular hypotheses researchers evaluate in the study.

Productive scientific problems and questions are those on which research likely will likely improve understanding substantially. Productive scientific problems are not necessarily popular ones, and studying them might not enhance your professional career.

Sometimes researchers have little to no discretion on the problems and questions they study, which are instead determined by their employers or clients. In such cases, the scientific productivity of the problems and questions may not be relevant. Other researchers may feel as if they have limited choice of research problems and questions, restricted by disciplinary boundaries and traditions, available funding, organizational priorities, peer opinions, and similar constraints. My advice applies to those willing and able to make scientific productivity a more important factor than these other concerns. Of course, if you are doing unfunded or self-funded research (see Chapter 16), many of the limitations to choosing problems and questions are absent.

Researchers-in-training and scientists early in their careers sometimes worry about being able to come up with interesting, relevant, and practical research ideas. Dreaming up original scientific problems actually is not important to being a good scientist. By building on or responding to prior work, researchers can find ready-made research problems and questions.

Interest

> To the natural philosopher there is no natural object unimportant or trifling. From the least of nature's works he may learn the greatest lessons.
> —John Herschel, 1831

> ... everything is interesting if you go into it deeply enough.
> —Richard Feynman, 1992

We scientists are inquisitive. If you choose a productive scientific problem or question, your interest in it will almost certainly grow as you study it. So don't despair if you lack a burning interest at the beginning. Also don't misinterpret your current lack of skill and knowledge related to it as a lack of interest. You can learn new skills and knowledge. However, if you are not interested enough to commit to completing a study, pick another problem or question.

Possibility of empirical study

One characteristic of a productive scientific problem is that researchers can study it empirically and readily. Data are the main feedstock for scientific progress. Even studies that generate relatively small amounts of data can be valuable. If you can do these studies fairly easily and quickly, you can learn at a fast pace and accumulate considerable and diverse data over time.

When there are very limited existing data on a problem, descriptions are inherently uncertain and rival hypotheses are difficult to evaluate, because their predictions might not be distinguishable empirically. Stalemates also occur when there are many existing data that researchers have analyzed repeatedly, and new data on the problem accrue slowly, as in some kinds of observational research. There may be several viable hypotheses and theories about the problem, but given the slow trickle of fresh evidence, it may take several generations before scientists can test their relative merits.

Research that is never completed (or never even begun) is by definition unproductive. Therefore, avoid studying problems that require massive ongoing funding, long periods of preparation, or difficult to access research sites, materials, or equipment. Many scientists have essentially wasted their careers by planning for events outside of their control to occur. They gambled that their investments of time and effort would pay off with major scientific advances from the ambitious projects. In many fields, long-term studies are potentially very valuable, but are also at high risk for ending prematurely. If you do pursue them, limit your potential losses by designing simple (see Chapter 13), low-cost projects that yield useful results on short-term outcomes and other problems along the way.

Controversies

Try to learn what people are discussing nowadays in science. Find out where difficulties arise, and take an interest in disagreements. These are the questions which you should take up. … pick up, and try to continue, a line of inquiry which has the whole background of the earlier development of science behind it … we cannot start afresh … we must make use of what people before us have done in science.
—Karl Popper, 1948

The question of doubt and uncertainty is what is necessary to begin; for if you already know the answer there is no need to gather any evidence about it.
—Richard Feynman, 1964

Problems or questions marked by debates and competing hypotheses are ripe for research if you can study them empirically and readily. Such problems and questions are usually well-defined, with the points of scientists' disagreements highlighting the uncertainties. Controversies involve other researchers who are highly interested in the problem or question, and therefore serve as eager critics of new work on the topic. This means you are likely to receive useful feedback on your study, especially if seek it (see Chapters 31 and 40). Research on controversies may also lead you into active debates with other researchers, so be prepared for possible drama and tension with them (see Chapter 43). You can find problems and questions marked by disputes by reading the literature carefully (see Chapter 47), noting disagreements that your colleagues have in informal discussions, and observing conference presentations, especially the question and answer sessions that follow them.

Problem context

Scientists sometimes study problems, especially applied ones, that end up being irrelevant in practice. Although the research may be sound, the original problem evaporates because research on another related problem covers or eliminates the original problem. For instance, knowledge of how to prevent a phenomenon, such as a disease, may make attempts to understand how to respond or treat it unnecessary. Of course, many phenomena cannot be prevented by human intervention, so it is not always possible to judge whether research on response-oriented problems will be displaced by research on related prevention-oriented problems.

Conversely, successful research and development on one applied problem can lead to new practical problems that must be addressed to realize the benefits of the original research effort. For example, a new effective medical treatment may involve significant side effects, canceling its therapeutic value unless researchers can develop good countermeasures. Technological research of many kinds is particularly susceptible to this dynamic. Sometimes, however, it is not possible to anticipate the likely negative effects of research and development on applied problems.

Furthermore, research on a narrowly defined specific problem may get supplanted by research on a more generally defined problem that includes the specific problem as a special case and has many scientific and practical applications. Therefore, think carefully and broadly about the context of the problems you might study, and consider their potential overarching, antecedent, and consequent problems before committing to any particular one (see Chapter 57).

Foundational beliefs

... the doctrines which best repay critical examination are those which for the longest period have remained unquestioned.
—Alfred North Whitehead, 1931

... axioms provide the foundations of a science and convenient points of reference, but they are not tablets of stone, and every scientist has the right to challenge them.
—Donald Braben, 1994

Few intellectual tyrannies can be more recalcitrant than the truths that everybody knows and nearly no one can defend with any decent data (for who needs proof of anything so obvious). And few intellectual activities can be more salutary than attempts to find out whether these rocks of ages might crumble at the slightest tap of an informational hammer.
—Stephen Jay Gould, 1996

The received wisdom in many, perhaps most, fields is fragile, built not on foundations of evidence and rigorous evaluation, but unsubstantiated belief and assumption. Usually, scientists do not discuss or even consider these beliefs and assumptions, and they may not even be aware of them. Examining assumptions requires questioning the basis for your work and field (see Chapter 61). Scientists often have not evaluated these underlying beliefs and assumptions empirically or thoroughly. In some cases, the assumptions cannot be evaluated empirically or are matters of definition, such that they are valid to the extent one agrees they are valid. However, when the assumptions and beliefs can be evaluated empirically, they are perhaps the most productive of all scientific problems and questions. Such work can indicate how sound the basis of research in a field is, point to very different directions for study, and result in major discoveries.

Research on foundational beliefs is unpopular and most scientists in a field oppose it violently. The social dynamics involved are an extreme case of those in controversies. The key difference is that you may be the only one to represent a side in a debate, with most researchers in the field on the other side. Bolster yourself with strategies to endure the opposition (see Chapter 62). Fortunately, the opposition is very valuable, as the criticism you receive allows you to improve the quality of your research.

Freedom from ideological and financial commitments

We scientists need to be passionately dispassionate about our research to do it well. This is difficult enough to approach, irrespective of research topic, with our psychological and social frailties as humans. It becomes nearly impossible to approach when we study problems related to ideas outside of science to which we are firmly attached. Religious and political ideologies and financial/commercial interests are classic pre-established biases that can harm our research on directly or indirectly connected problems. Similarly, personal or professional commitments to (or hostility toward) particular interventions, views, and policies can warp our pursuit of truth on related scientific problems. Even deeply held beliefs about human nature, justice, destiny, and other philosophical concepts can distort our work. It is especially dangerous when researchers are motivated to study a particular problem to obtain confirming proof of a related non-scientific belief. In all of these circumstances, scientists are likely to protect their prejudices, perhaps unwittingly, by designing research poorly, misinterpreting results, ignoring counterevidence, and dismissing constructive criticism.

How can you tell whether your beliefs may prevent you from studying a problem rigorously and well? One test is to ask yourself whether there is any evidence that could lead you to question or change those beliefs. Can you imagine having very different, even opposite, beliefs? If you answer no, then avoid studying problems that are even tangentially related to those beliefs.

Financial commitments are generally less problematic than ideological ones. Most commercial products and services face independent evaluation by the public in the market, with corresponding financial rewards or penalties. Many face regulatory scrutiny by government agencies. And those who sell goods or services also are legally responsible for them and claims they make about them. These kinds of checks are far from perfect, but they do reduce bias in commercially-related research. Ideologically-driven researchers encounter no such controls. Furthermore, current norms about researchers disclosing conflicts of interest focus on financial conflicts involving private companies and ventures only. Thus, ideological conflicts of interest are insidious and may be pervasive in many fields.

Substituting one problem for another

Sometimes researchers aim to study one problem but actually study another several steps removed from the one they intend. For instance, they seek to understand human biology and behavior but study a model organism. Another example is when scientists try to learn about the empirical world via simulation. Or they may use proxy measures and artificial settings with poor or unknown correspondences to the outcomes and settings of real interest. In all such cases, scientists can advance understanding, but only on the proximate problems they truly study. However, their work may be necessary for and inform — through hypothesis and speculation — later research on the ultimate problem of interest. The key is not to fool yourself or others on what problem you are really studying.

Problems and questions to avoid

If you pick a problem or question based on the availability of or preference for a method, intervention, or data set (see Chapter 50), your research is at risk of scientific failure or irrelevance. Also, when existing research on a problem or question has significant methodological or design limitations, additional research with the same methods and designs is unlikely to advance understanding. It is better to implement or develop stronger approaches, even if it involves much more time and effort (see Chapters 3 and 52).

Process for choosing problems and questions

Economic principles can help in selecting scientific problems and questions. Consider the *opportunity cost* of studying one problem or question rather than another. Time and effort you spend on one is time and effort that you cannot spend on another, so weigh your options thoughtfully. Moreover, estimate the *profit*, in a scientific sense, that might come from research on a problem, relative to the effort, time, and cost required to do it. You may decide that the potential scientific benefit is just not worth your investment, no matter what your alternatives are.

Search the literature thoroughly (see Chapter 5) as you develop a research question. Such searching can correct any misperceptions and misunderstandings you have and keep you from doing irrelevant or wrong-headed research. Your search will also indicate how your question relates to prior work. If your research question has been studied before, don't give up. You can still study it. Just be sure that at least some parts of your research will be comparable to the prior work so that you build knowledge (see Chapter 14). Studying the same question as others, but in different ways also is valuable.

49

Crossing Disciplinary Boundaries

Everyone follows his own path. Some are extensively prepared and follow the furrow that has been traced out, etc. For myself, I arrived on the scientific field by devious paths, and I freed myself of the rules by cutting across the field, which others would not perhaps have dared to do. But I believe that ... this has not been bad, because it led me to new vistas.
—*Claude Bernard, 1850–1860*

What one fool can do, another can.
—*Silvanus Phillips Thompson, 1910*

... the appeal to authority of experts should be neither excused nor defended. It should, on the contrary, be recognized for what it is — an intellectual fashion — and it should be attacked by a frank acknowledgement of how little we know, and how much that little is due to people who have worked in many fields at the same time.
—*Karl Popper, 1993*

Crossing borders between fields is vital for science. Most problems do not fall neatly within the somewhat arbitrary silos of academic fields. These fields arose in part due to social, political, and economic factors, and to some extent, the precise divisions are accidents of history. Discipline-bound research usually is incomplete and sterile, rarely producing great insights or advances in understanding. Without fertilization by other fields, a discipline becomes stagnant.

Working in a field new to you can also be exciting, as your pace of learning is faster than in fields in which you have studied longer and have less to learn. Interdisciplinary research keeps scientists sharp.

The best scientists are those who work in multiple fields and bring their diverse skills, knowledge, and experience to bear on the problems they study. Whatever they may lack in specialization is outweighed by these assets, which tend to confer outlooks and approaches for solving problems successfully. Interdisciplinary researchers also may be less submissive to the leaders in a field and the reigning theories, hypotheses, and assumptions.

There are several requirements for crossing disciplinary boundaries effectively. First, you must have a keen interest in a topic that has multidisciplinary aspects. This motivation will carry you as you face the challenges of working in a different field.

Second, working in a field new to you involves extra work. Prepare to feel dumb and confused while you strive to gain a basic understanding. Your ignorance, however, also is a strength —it allows you to see the territory with a fresh perspective, almost that of a child, not burdened by convention and ideology from formal training in the field.

Third, summon confidence that you can contribute on this new topic, no matter your prior experience and training. Scientists, especially in academia, irrationally dwell on irrelevant degrees and intellectual pedigrees, and often are hostile to those they consider outsiders or interlopers. Bolster yourself against such tribal attacks and focus on doing research. As a scientist, you already know the general logic, principles, and practices of research (covered partially in this book) and this knowledge serves as your passport. In 1267, Roger Bacon recognized mathematics as "the gateway and key to all other sciences." An understanding of basic mathematics and statistics also is essential for venturing into other fields, as is a willingness to increase your knowledge of these diverse tools.

Finally, cross-disciplinary researchers need to pay attention to the history of research on the topics they study. Some would-be interdisciplinary researchers act as if they have made significant discoveries or solved the problems at hand, when in fact, they have not contributed anything new or meaningful. Beware of the "Tarzan complex"—thinking that you can become an instant expert on a subject just because you have mastered another. This attitude will not lead to good scientific research, and is sure to alienate other researchers working on the subject. Crossing disciplinary boundaries means joining a conversation in progress, and it pays to know what others have already said (and found) (see Chapters 5 and 47).

Many interdisciplinary research projects involve specialists from different fields, but such representation is no guarantee of success. The critical ingredient is that the views, methods, and approaches from different fields get mixed together *in individual researchers*. Such learning occurs in the context of long-term and friendly collaborations between scientists in different fields, or when individual researchers immerse themselves in new fields. Good interdisciplinary research involves not a rigid division of labor by discipline, but a true meeting of minds and ideas.

50

Abolishing the Law of the Method

Give a small boy a hammer, and he will find that everything he encounters needs pounding.
—*Abraham Kaplan, 1964*

It is dangerous to submit to the "law of the method" (others have called it "law of the instrument" or the "golden hammer"). The law refers to the unthinking application of a particular study design, field site, population, or method of data collection or analysis to a seemingly unlimited range of scientific problems, even though doing so may be inappropriate or nonsensical. A variant of the law is applying a particular intervention, such as a treatment or technology, indiscriminately to one or more practical problems—a solution in search of problems.

In modern science, researchers often must specialize to a high degree. To learn a method well or develop an intervention often requires considerable time and effort. Thus, a scientist can be tempted to extend a favored method or intervention to as many problems as possible to increase his or her apparent scientific value (by being able to work on diverse problems) with minimal extra cost in labor or time. At its extreme, the law of the method can produce trivial variations on a research theme, as researchers attempt to get the most out of the approach, evidently because they have not identified more meaningful applications.

The law of the method also arises in fads that run through a field. Researchers often fall in love with particular new techniques, pieces of equipment, or software, viewing them as badges of quality or prestige, just like fancy consumer goods. Sometimes, researchers are determined to use the trendy method even when it is irrelevant, harmful, or expensive to their projects.

There are several options for abolishing the law. Spend the time and effort to learn new methods appropriate for the problems you want to study. You can also work with other scientists who already have mastered the methods you need. Or simply content yourself with studying just those meaningful problems for which your method is well-suited.

However, how can you even know that you are submitting to the law of the method? Scientific modesty is always a good goal, but far from a reliable way to help in this situation. One symptom of submission might be looking for problems for which the method could be applied. By selecting the method before the problem, you increase the risk of extending the method inappropriately. Good scientific research is *problem-driven*, not method-, design-, or intervention-driven. Modesty coupled with curiosity about methods used in other fields (see Chapter 47) may also help prevent applying a method to different problems blindly. Experience in other fields may demonstrate more appropriate methods for particular problems or at least lead you to search for relevant methods in other fields. Moreover, get feedback on your idea of using a method for a particular problem from researchers who use different methods for that problem. These other researchers could be blinded by the law of the method as well, but their criticisms of your method for the problem may still be good to consider.

51

Hypothesis Evaluation,
Description, and Exploration

We have a way of checking whether an idea is correct or not that has nothing to do with where it came from. We simply test it against observation. So in science we are not interested in where an idea comes from.
—*Richard Feynman, 1963*

... there is no sense in calculating the probability or the chance that something happens after it happens.
—*Richard Feynman, 1963*

Hypothesis evaluation, description, and exploration are three activities that together cover most empirical scientific research (see the Notes for the exceptions). Yet many scientists are confused about the differences between these distinct activities. It is crucial to know what kind of research you are doing, as the type determines the meaning and potential impact of your work.

Hypothesis evaluation

An hypothesis is simply a guess about a specific phenomenon that can be evaluated empirically. As Feynman emphasized, the origin of an hypothesis does not matter. Theories, results from prior research, speculation, and insights are all suitable sources, and none is better than another. Induction is just as good a path to an hypothesis as deduction. Some scientists insist that hypotheses and research be theory-driven. However, existing theories may be weak, incomplete, or irrelevant to the topic under study. Moreover, science would rarely advance if researchers restricted their future research to established past ideas.

In hypothesis evaluation, by definition, the hypothesis comes before the evaluation. All experiments are hypothesis evaluations, at least with respect to the manipulated variable(s), which must be set before designing the study and collecting the data. Every true hypothesis evaluation can be pre-registered (see Chapter 4). Scientists who pre-register their hypothesis evaluations show that their hypotheses did, in fact, precede their evaluations.

Hypothesis evaluation is not the same as statistical hypothesis testing. The latter is an approach to data analysis that is generally uninformative (see Chapter 8) and often applied, incorrectly, to research that is not hypothesis evaluation.

Hypothesis evaluation is fun. There is nothing like making an expectation explicit and then seeing what happens. It doesn't matter what the outcomes are. Hypotheses capture our attention and increase our anticipation of the results. Evaluating specific ideas empirically produces a tangible feeling of progress that is thrilling.

Description

Description involves collecting, analyzing, and reporting *all* data related to a particular focus of study. Targets of study can vary dramatically within and between fields, and can include some kind of specimen, artifact, location, area, material, substance, organism, behavior, disease, community, belief, event, process, structure, sequence, or system, among many other possibilities. Sometimes the goal of description is to estimate a single quantity. In description, the goal is to provide a data set or summary that details the focus of study rather than test an hypothesis. Computing summary statistics for all variables under study and, perhaps, associations among them (or a predefined subset) also is a kind of description. Fitting a model to data is another type of description (if you did not specify the model as an hypothesis in advance), as the model is just a summary of the data.

Often we scientists know exactly what we want to study in a systematic way, but do not have enough prior knowledge or insight to venture an hypothesis. Such research is description. Description is just as valuable in scientific terms as hypothesis evaluation. In many fields, descriptive research is the main source of facts and understanding. Some researchers feel compelled to make several hypotheses for each study, even when they do not have good reasons to expect a particular result. Such hypotheses are literally pointless. It is fine to identify what you will study in a descriptive fashion without committing yourself to an hypothesis. You do not need an explicit hypothesis as an excuse to study something.

Pre-registered protocols for descriptive research are still important for defining the target of study, the characteristics to be described, and the methods to be used. The pre-registered protocols serve as benchmarks for assessing how completely the researchers report their results and proof that the results reflect a systematic plan. Incomplete reporting could indicate that the researchers are suppressing results.

We can also evaluate hypotheses with and generate hypotheses from the results of descriptive research. Sometimes the published results have direct implications, without additional analysis, for one or more hypotheses. In other cases, the data and/or results from the descriptive research can be analyzed further for hypothesis evaluation. Large and complex descriptive data sets, in particular, are useful, as inspecting them casually may not yield any insights into patterns. Thus, researchers can generate hypotheses with little threat of prejudice and then evaluate them with the data.

Exploration

Exploration involves identifying phenomena through informal search or happenstance. It entails no objective, definable procedures specified in advance. As exploratory research proceeds, decisions for further data collection or analysis are contingent on data already reviewed, analyses already done, or the whim of the researcher.

Exploration in data analysis can take many forms. Hunting for interesting results in a mass of data (data "dredging," "snooping," "fishing," or "mining"), manipulating the data or analyses to get "statistically significant" results ("p hacking"), and developing hypotheses after the results are known ("HARKing", post hoc analysis) all are exploratory activities. Although researchers often compute statistical hypothesis tests for exploratory data analyses, they are invalid, as Feynman noted, because the researchers formed or selected the so-called hypothesis after reviewing the data and/or analyses.

Non-scientists can engage in exploration, as it sometimes requires no special training. For instance, some lay persons believed that images from the 1976 Viking orbiter mission to Mars showed a mountain that looked like a face. They interpreted this as an artificial structure, possible evidence of a past Martian civilization. More recent orbiter images show that the appearance of a face in the earlier images was due to low resolution and lighting effects. Many scientists scoff at such cases as poor reasoning by non-scientists. However, this kind of exploration is logically no different than professional exploratory research in many fields, including planetary science and astronomy.

A result from exploratory research is equivalent to an hypothesis. It is no more valid than an hypothesis derived by other means. That can be a humbling thought if a project took a lot of time, effort, and money to carry out and the researchers believed their work would amount to more than an hypothesis.

Useful applications

Exploration can be a reasonable approach when there is little to no prior research on the topic. It can help generate topics for study, research questions, and hypotheses, and inform decisions about designs, methods, and logistics of future systematic research. Exploration often takes place during informal experiences and observations in the field (see Chapter 2). Such experiences and observations often may not be worthy of publication, given their unsystematic nature, but they can guide subsequent research. Exploration, though, isn't always the only option in these circumstances. Often you can still plan in advance some data to be collected and analyses to run for descriptive purposes in anticipation of opportunities present in similar contexts as the one to be studied. Even if some of these opportunities do not occur, you will be prepared for those that do.

Exploration can also be useful when researchers evaluate an hypothesis and get unexpected results (usually contrary to the hypothesis). The researchers then perform additional analyses to explain the finding. They do not plan these analyses before data collection or even plan them before running them. These analyses typically are not guided by any logic other than an attempt to see whether the unexpected result can be accounted for by other factors. It is almost like a detective trying to solve a criminal case, based on intuition and guessing, ignoring some clues but investigating others in idiosyncratic fashion.

Exploratory data analysis is undoubtedly fun to do. It *feels* like your understanding is growing from the exploration. But this feeling is an illusion. Unplanned analyses can only lead to ideas and hypotheses, not evidence.

Exploration also can be helpful as a learning tool. It gives researchers firsthand data analysis experience with various statistical phenomena, such as apparent confounding, moderating relationships, and artifacts. This kind of experience may help researchers develop skills in anticipating potential alternate explanations to observed results. Such skills can then be applied when creating study protocols to be pre-registered. Furthermore, exploratory data analysis can teach scientists lessons about their susceptibility to wrong interpretations, as when exploratory analyses are not replicated in other data sets.

Problem

The greatest problem with exploration is that many researchers do not recognize its limitations. In some fields, many researchers think they are doing hypothesis evaluation when in fact they are doing exploration, and they misrepresent their work unknowingly. This is one reason why research often is not reproducible. Exploratory research may even predominate in these fields. This leads to massive diversion of effort and funding to a relatively unproductive activity that yields little reliable scientific knowledge.

Classifying research

Any study or report can include combinations of hypothesis evaluation, description, and/or exploration. Usually it is easy to classify a particular research activity. However, there are some kinds of research that are or seem hard to classify.

Analysis of previously analyzed data can be any of the three main empirical research activities, depending on the circumstances. Published results of the original analyses or direct inspection of the data can prejudice researchers planning to reanalyze the data. If the published results or data inspection inform hypotheses the researchers hope to evaluate, then the reanalysis is exploration, not hypothesis evaluation. If the reanalysis plan is not guided by prior results or data inspection, then the research can be either hypothesis evaluation or description, as the case may be.

Sometimes researchers divide a data set into halves, often on a random basis. With the first half, they conduct descriptive or exploratory analyses, whether manually or with "black box" machine learning methods (see Chapter 55), and then arrive at a particular result as their final choice. This is exploration. With the second half, they attempt to validate the result from the exploratory first half—that is, to check how similar the results are between halves. This validation phase is a narrow kind of hypothesis evaluation. It is narrow because the hypothesis and evaluation data come from the same study and data set, and thus share many of the same patterns, errors, and biases.

Systematic reviews and meta-analyses (see Chapter 15) can be hypothesis evaluation or description, and are classified by the same criteria as primary research. Unsystematic reviews, however, often are exploratory efforts. It is possible to synthesize results from hypothesis evaluation, description, and exploration together in a systematic review, although exploratory results tend to be the least reliable and most muddy the literature. In practice, many systematic reviews include results from multiple types of research, and it usually is not possible to classify included studies with confidence based on the available information.

Ironically, many scientific expeditions to locations not previously visited by people—such as other planets, ocean deeps, or remote lands—are typically focused on description, not exploration, in the scientific sense. For such projects, researchers design and plan their studies in advance. Their instruments and measures are pre-determined, and their goal is typically to measure as comprehensively as possible. Specific sampling decisions in these projects tend to be made in an ongoing fashion. Exploration is probably most prominent in researchers' scanning the scene, and the information they gain in such scans informs sampling decisions for their descriptive activities. The end results are descriptive, not exploratory, even though the effort may still produce discoveries of new phenomena. The descriptive data and results, of course, may also be used for subsequent hypothesis evaluation or exploration.

Authors' claims about the type of research they did may or may not be accurate. Moreover, authors often do not label their work explicitly as hypothesis evaluation, description, or exploration. Experiments are more likely to be hypothesis evaluation, although beware of analyses of outcomes that the researchers did not hypothesize initially to be affected by the experimental intervention. Replication studies are also not likely to be exploration. Of course, many researchers in the past actually did hypothesis evaluation or description, and avoided exploration in their work, but just did not give proof of it.

On a philosophical level, descriptive and exploratory research do involve hypotheses of a sort. The measures we use and observations we make stem from what we think *might* be relevant. These thoughts are effectively unstated, unelaborated potential hypotheses. We may not even recognize them as such. But by necessity, we must make what amount to theoretical decisions *before* we design and conduct research of any kind. However, on a practical level, we do not evaluate any hypotheses in description or exploration.

In your reports, make clear what is hypothesis evaluation, description, and exploration. Pre-register your research so that others can know with certainty in what category your work falls.

52

No Substitute for Experimental Control

... do thou experiment so that you mayest acquire knowledge. Scientists delight not in abundance of material; they rejoice only in the excellence of their experimental methods.
—*Jabir ibn Hayyan, A.D. 721–815*

In the sciences, there is doubtless a very close connection between observation and experimentation. Nevertheless, it is necessary to distinguish them because (otherwise) everything would become confused.
—*Claude Bernard, 1850–1860*

The words "experience" and "experiment" have a common Latin root. The wisdom of experience comes from *doing* as well as *watching*. Our key insights originate from interacting with—in scientific research, we say intervening with or manipulating— our surroundings. In the strictest scientific sense, experiments are systematic studies in which researchers not only observe but also manipulate nature. Typically, scientists try to produce a hypothesized cause and then observe what happens, looking especially for the hypothesized effect.

Experiments are for evaluating hypotheses about causes and effects (see Chapter 51). Many scientific research questions are not about causal relationships, and therefore experiments are inappropriate or irrelevant in those contexts.

Experimental control

Experimental control has two essential aspects. First, researchers intervene in a certain, *standardized* way under particular conditions. Second, researchers *compare* what happens with the intervention to what happens when there is no or a different intervention. Both aspects of experimental control are necessary for making causal inferences.

Often, study units (such as individual organisms, persons, or places) vary from instance to instance or can change over the duration of the experiment. In these circumstances, researchers compare a group (for example, of individual organisms, persons, or places) that gets the intervention (the experimental or treatment group) with a group that gets no intervention (the control group) or a different intervention. To make sure the groups are otherwise similar, researchers usually assign units to groups randomly (by some chance mechanism) or according to a balanced, systematic pattern.

Sometimes, natural experiments occur in which haphazard events not produced by the researchers serve as the experimental manipulation. Studies of such episodes can approach the methodological rigor of true experiments, but the degree of rigor depends on the details.

Observational research

In observational research, scientists do not intervene with nature, but simply observe it. We humans tend to see what we want to see in observational research results. Correlations seem to indicate causes, and alternative explanations can be difficult to muster. In many fields, researchers use longitudinal assessments, measure covariates, make assumptions, and apply analytic techniques to attempt causal inferences from observational data. Likewise, researchers in many fields apply several criteria for inferring causal relationships from observational evidence. However, no such procedures or criteria for observational research can deliver the unique strength of experimental evidence: elimination of every other potential cause apart from the hypothesized one. With observational evidence, there are *always* other factors, including those researchers have yet to imagine, that have not been assessed in relation to the hypothesized cause and effect. Only in experiments can researchers manipulate the hypothesized cause and control all other factors, known and unknown.

The difference between experimental and observational evidence is logical and qualitative. No amount of observational evidence can equal in explanatory weight some experimental evidence. That is, a lot of weak evidence about a causal relationship still adds up to weak evidence. Scientists often fail to infer causes from observational evidence correctly, when judged by the results of later experimental research and experience.

Scientists can use observational research to evaluate hypothesized causal relationships in limited ways only. For example, if two variables are related causally, they should tend to be correlated in observational studies, and the hypothesized cause should tend to precede the hypothesized effect. Of course, variables not related causally can show the same patterns (see Chapter 56). Observational evidence also can be a check on experimental evidence in one specific scenario. Causal relationships that scientists find in experimental research are suspect if they are inconsistent with the observational evidence. Furthermore, researchers can use observational results consistent with a particular causal hypothesis as a warrant for evaluating the hypothesis experimentally.

There are many successful scientific fields built almost entirely on observational evidence, such as geology and astronomy. Interestingly, these successful observational fields are anchored in experimental knowledge derived from other fields such as physics and chemistry. Nonetheless, causal knowledge in such fields is *inherently* less certain than fields in which evidence is primarily experimental. Causal knowledge is even more uncertain in observational fields that lack anchoring in experimental knowledge.

Experimental context

Researchers often conduct experiments in artificial conditions where they can establish experimental control easily, such as in vitro, model organisms, or laboratory environments. These settings may be quite different from the field contexts of ultimate scientific interest, such as in human patients or natural conditions (in the wild or community). Thus, results from experiments in artificial circumstances are nothing more than *hypotheses* about what occurs in the field naturally.

Experiments in artificial settings are often precursors to experiments in the field, which tend to be more costly, difficult, and error-prone. Nonetheless, move experiments to the field or clinic as quickly as possible. It's easy to spend many years and huge sums of money on laboratory experiments with results that cannot be reproduced in the field. Similarly, question lines of experimental research in artificial settings that have little relevance to or good analogs in natural conditions.

Experimental interventions, whether evaluated in artificial or natural settings, always are restricted in time and space. No matter how strong or consistent experimental results are, an intervention may not achieve the same effects when implemented broadly or universally, such as throughout a population or ecosystem. A persistent and pervasive intervention may change the underlying dynamic, possibly even producing effects opposite from those found in experiments. Applying experimental knowledge to real-world practice thus can involve a fair amount of extrapolation and associated uncertainty.

Sometimes researchers do experiments in settings (whether in the laboratory, clinic, or field) or with procedures that are very favorable to producing a large effect. Results from such experiments may overstate the strength of a causal relationship if the experimental conditions are not typical of the circumstances to which the researchers try to generalize.

We can study phenomena that occurred in the past or may occur in the future experimentally only by extrapolating experimental knowledge from the present. However, we and future scientists can replicate prior experiments to assess how much causal relationships have changed (or not) over time.

Debasing experimental control

Specific features of experimental designs and methods for limiting bias vary by field and application. Experiments that lack these features produce lower quality evidence than experiments with these features. In some cases, the lack of bias-reducing features may lower the quality of the experimental evidence to a level not much higher than observational evidence, although this may be a matter of judgment. Strive to implement all relevant bias-reducing measures in your experiments. Experiments may also be debased just like observational studies when researchers do not follow the other essentials for scientific research I describe in this book.

In some fields, researchers do experiments in which the intervention is really multiple interventions in one. That is, the intervention has multiple components, each typically focused on a different hypothesized cause of an effect. Such experiments are of limited scientific value because it is impossible to know which particular component (and corresponding cause) or combination of components led to a specific effect. Interventions with multiple components focused on applied problems also are unlikely to be implemented in practical settings given their greater complexity and cost. In this context, applied practitioners have no empirical basis for selecting components to use. To prevent these handicaps, design experiments in which each intervention component is a separate experimental condition, and compare them with the control condition and possibly one or more sets of components.

Scientists sometimes degrade the logical superiority of experimental evidence by transforming it into observational evidence when comparing results between experiments. Such comparisons are observational in nature, even though the underlying studies are experiments. An elaborate example of this is the currently popular technique of network meta-analysis. The causal inferences from such approaches are no more legitimate than those from other sorts of observational research. To address this problem, compare the particular interventions or conditions directly *within* new experiments.

In applied settings, researchers often do not evaluate interventions with full experimental control, even though it might be possible, due to political, economic, or ideological conflicts of interest. A proper experimental evaluation could show clear evidence that a favored intervention is not effective.

Prioritizing experiments

If you have a causal hypothesis and experiments are possible for your topic, go straight to true experiments. This is especially important when the outcomes—the effects—of interest take months, years, or decades to appear. A long program of observational research just wastes time, effort, and resources. Consider observational research designs only if experiments are not yet possible.

Many researchers in applied settings, especially those involving human participants, choose uncontrolled experiments —those without a control group—when first evaluating an intervention. The purpose an evaluation is to assess the effectiveness, safety, and other aspects of an intervention. Without a controlled true experiment, there is no good evidence on these outcomes. An intervention could be ineffective or harmful, but the results from an uncontrolled experiment would not necessarily reveal it. Therefore, it is unethical for researchers to implement an intervention in an evaluation without full experimental control.

Controlled experiments also are essential when first assessing an intervention because sometimes an intervention becomes popular among applied practitioners after only uncontrolled experiments. It then becomes politically or ethically impossible to withhold the intervention in future studies, as in controlled experiments, even though scientists have not yet established its merit rigorously.

First or early controlled true experiments can still be used to refine the intervention and gather other formative data. Observational studies can also occur within an experiment, such as watching what happens naturally within a control group. Moreover, true experiments are possible, in many circumstances, even when there is only a single case or unit under study.

Although experiments are powerful tools for studying cause and effect, no single experiment is ever definitive. Replication is as important for experiments as it is for observational studies, especially when they are based on small samples (see Chapter 14).

Once you have formed a causal hypothesis, experiment!

53

You are not Your Hypotheses or Results

... I wished it to be true; but, alas! A scientific man ought to have no wishes, no affections—a mere heart of stone.
—Charles Darwin, 1857

An experimenter must not hold to his idea, except as a means of inviting an answer from nature. But he must submit his idea to nature and be ready to abandon, to alter or to supplant it, in accordance with what he learns from observing the phenomena which he has induced ... Men who have excessive faith in their theories or ideas are not only ill prepared for making discoveries, they also make very poor observations. Of necessity, they observe with a preconceived idea, and when they devise an experiment, they can see, in its results, only a confirmation of their theory.
—Claude Bernard, 1865

... the vast majority of hypotheses prove to be wrong.
—W.I.B. Beveridge, 1950

You investigate for curiosity, because it is unknown, not because you know the answer.
—Richard Feynman, 1956

Refutations have often been regarded as establishing the failure of a scientist, or at least of his theory ... Every refutation should be regarded as a great success; not merely a success of the scientist who refuted the theory, but also of the scientist who created the refuted theory ....
—Karl Popper, 1962

If we identify with hypotheses, we are prone to evaluating them poorly. If we desire particular results or feel that the results somehow reflect on us as scientists, we are liable to influence or misrepresent findings.

It is impossible to learn anything new if we never refute hypotheses. We should hope that our expectations will often be wrong.

Although you are not your hypotheses or results, you are your designs, methods, and reasoning. That is, *what* you think and find do not define you as a scientist, but *how* you think and what you *do* do reflect your scientific skill. An hypothesis is just a guess and results are (usually) mostly a product of nature. In contrast, you control and are entirely responsible for the way you design, conduct, and interpret your study.

Any result from a study designed and conducted well is a good result. Understanding increases no matter what the outcome is.

Strategies for detachment

Several approaches may help you avoid identifying with your hypotheses and results. For instance, work on several problems with different hypotheses. You might be less tempted to identify with an hypothesis if your effort is divided than if you focus on one or two hypotheses or theories only. The potential feeling of loss from a refutation or undesirable result also might be less when you study many problems, hypotheses, and theories.

Study problems and questions with multiple competing hypotheses and evaluate as many of them as you can, simultaneously if possible. This shifts your focus to the quality of the evaluation rather than the fate of a specific hypothesis. In 1892, Thomas Chrowder Chamberlain noted that generating and evaluating multiple hypotheses for a given phenomenon has lasting intellectual benefits: "the mind appears to become possessed of the power of simultaneous vision from different points of view." This skill is essential for reasoning, doubting, and questioning (see Chapters 56, 57, and 61).

Evaluate hypotheses to which you have no pre-existing attachments, such as those proposed by others. Also, study problems with which you have no ideological, financial, or professional conflicts of interest (see Chapter 48).

It is also easier to ignore the appeal of an hypothesis by studying problems in which you have a personal stake in the truth. For example, in medical research, focus on a disease that you have or a loved one has. If your *only* concern is improving treatment or prevention of a disease, you are unlikely to fool yourself that a weak hypothesis is strong. When we scientists align our research agendas to find truth, we hold on to particular hypotheses and theories only to the degree they are consistent with the empirical evidence.

Even the words you use to describe hypotheses may help remind you to be detached. Avoid referring to a hypothesis, theory, or model as "my" or "our", even if you developed it. Give it a descriptive name instead. By objectifying and de-personalizing the idea you may be more willing to attack it and accept its demise if it fares poorly in empirical evaluation.

Likewise, do not describe hypotheses as being proved, supported, or confirmed in empirical tests, as they cannot ever be. In empirical evaluations, hypotheses can only survive or be falsified (or fall somewhere in between). Confirmatory terms are not just inaccurate (see Chapter 27), they also express false beliefs in hypotheses. As Karl Popper often noted, an hypothesis that survives an empirical test means nothing more than it is worthy of further empirical evaluation and criticism.

We scientists also can profit by taking an engineer's adversarial attitude toward hypotheses. Scientific hypotheses are tools for understanding nature. When engineers test their tools and other objects they design, they try to break them—to understand the circumstances in which they fail. Engineers then face the ultimate proving ground of how an object functions in the real world. If the testing procedures were lacking in any way, the object could fail in the real world, possibly with deadly consequences, and it would be the engineers' fault.

Most scientists lack such an ultimate proving ground, and very rarely shoulder such responsibility. Perhaps as a result, our testing—hypothesis evaluation—tends to be less adversarial. Indeed, engineers who design and build an object may be the strongest advocates for especially rigorous testing of it. We scientists would do well to imitate them.

54

Simulations are ... Simulations

Supposing is good, but finding out is better.
—*Mark Twain, 1835–1910*

The skeptic will say: 'It may well be true that this system of equations is reasonable from a logical standpoint. But this does not prove that it corresponds to nature.' You are right, dear skeptic. Experience alone can decide on truth.
—*Albert Einstein, 1950*

A danger ... is the temptation to substitute simulation for its own sake for the appropriate subject ... in the natural world.
—*Philip Anderson, 2011*

Simulations, mathematical modeling, and computer modeling generally refer to the same or nearly the same activity: using mathematics and computerized calculations to represent some sort of process or system. Computer models are powerful tools, and thus present both opportunities and risks.

Models of all types are essentially theories or hypotheses. They focus on description and prediction. Descriptive models are summaries or abstract representations of empirical phenomena. (Models can be developed from empirical research based on description, but descriptive research is typically not driven by a model; see Chapter 51.) Predictive models can be used to produce hypotheses—often quite specific, quantitative ones—for new empirical results. Frequently, models are both descriptive and predictive.

Computer models offer some advantages over other kinds of models. As virtual sandboxes or toy universes, they allow researchers to specify their ideas formally and observe easily the interactions and implications of those ideas as if in a vacuum, isolated from other influences. Simulations can also suggest possibilities about the topic under study that scientists might not have otherwise imagined.

As commonly implemented, however, computer models can be problematic for description and prediction of empirical phenomena. Although not necessarily specific to computer models, the following aspects of model development and testing are important to consider.

Comparison with empirical results

No matter how many simulations have been run, a model is still just an idea (or set of ideas) until researchers compare its output with empirical results. Sometimes scientists report their modeling exercises without any comparison with empirical data. These reports are logically no different and no more valid than any other statement of an hypothesis or theory. Combining several models into ensembles does not increase validity; rather, it simply compounds hypotheses that are often not logically independent of each other. Also, comparing a model with empirical observations is meaningful only if the researchers did not rely on the observations in any way in developing the model (see Chapter 58).

Unmeasurable parameters

Some models include parameters which cannot be measured well or at all empirically. This makes these components of the model unfalsifiable and limits the descriptive value of the model. Even if the simulation results correspond well with empirical results, such a model may be right for the wrong reasons.

Limited validation data

All types of modeling efforts are hampered when the empirical validation (comparison) data are limited. In some fields, there may be hundreds or even thousands of models but only one or very few data sets for comparison. If the topic makes it difficult to collect new data, or new data accrue very slowly (such as with time series), even the best models are at high risk of overfitting, tuned to particular aspects of the limited data that are not necessarily general to the phenomenon under study. Validation data sets with few observations worsen these problems. The most extreme situation is when there is only one observed instance of the phenomenon under study, such as the Earth or the universe. In these cases, there may be many models consistent with the single observed instance, with no good way to judge their relative merits.

Empirical inputs

Many models involve using empirically-based parameter estimates. The best empirical estimates come from well-done meta-analyses and systematic reviews (see Chapter 15). A model inherits all of the defects of these empirical estimates. If these estimates are biased, or the parameters are actually irrelevant in the modeled phenomenon, the model output will be meaningless (the "garbage in, garbage out" principle).

Sensitivity

Simulations have a start and a finish. Modeling results can be quite variable depending on parameters' initial values. Likewise, simulations relying on single values or restricted ranges of parameter estimates can produce artifactually narrow ranges of model results. Simulations across a broad range of possible parameter values—to account for uncertainty—increase the chance that model output will reflect at least some realistic scenarios. Similarly, modelers can investigate the most critical aspects of their models by noting the effects of adding or removing parameters and altering details of the model's structure. As models become more complex, it becomes increasingly difficult to evaluate their sensitivity in all of these ways, especially in parallel and with consideration to the interactive (combinatorial) possibilities. This is one practical reason why simple models are best (see Chapter 13).

Transparency

Many modeling exercises represent feedback loops—developing the model, then comparing outputs with empirical data, then modifying the model, then comparing with data again, and so on. Modelers often do not report the full development, modification, and interim comparison process, or the rationale for each modification. Without a description of these details, readers may rightfully assume that the model is ad hoc and subjectively determined. Moreover, unless the precise features of a predictive model are specified prior to any examination of the empirical comparison data, model evaluation actually amounts to exploratory research—a statement of an hypothesis, not evaluation of one (see Chapters 4 and 51).

Application

Modelers often advocate for using their models to make decisions in the practical realm outside of science. Sometimes modelers make this argument even when the model has not been compared with empirical data or suffers from other problems I have highlighted here. Model estimates are not facts. The history of science shows the impermanence and insufficiency of specific models. Unless a model has a long and solid track record in description and prediction, it is foolish to rely on it as the basis for important decisions in the real world.

Models are essential for scientific progress. Computer modeling has the potential to speed advances, but also to distract and divert us from accumulating reliable knowledge. By attending to the aspects of modeling in this chapter, we can better discern in which direction particular modeling efforts are leading.

55

Analysis for Understanding

Although they were able to make forecasts of great accuracy, they did so in a way which did nothing to explain the events in question. Their work made eclipses, conjunctions, and retrogradations predictable, but it made them no more intelligible than before. One could go on thinking of the planets in any way one pleased, and the regularities in their motions remained fundamentally mysterious.
—Stephen Toulmin and June Goodfield, 1961

Across millennia and cultures, humans have studied nature, seeking to identify patterns in weather, behavior and development of organisms, movements of heavenly objects, and other phenomena. These efforts have often involved intensive observation and careful calculation. They have also produced highly accurate predictions and classifications that have served as useful knowledge. Yet until some people started studying nature in a self-consciously scientific way, this knowledge did not include, for the most part, understanding. Despite the rigor of their underlying observations and calculations, people in the pre-scientific era generally explained natural phenomena by invoking supernatural forces or creating illogical stories.

Similar strains of this approach to research—analysis without understanding—persist in modern science. Classification and prediction are central activities in scientific research, but classification and prediction alone do not constitute science, no more than accurate forecasting of the solstice amounts to a theory or description of Earth's orbit and tilt. Scientific understanding is about *how* and *why* things happen. Many currently popular analytic approaches in some fields, such as machine learning and related techniques, are "black boxes": they may produce practically useful classifications or predictions but no insight to the specific factors that might underlie those classifications and predictions. Even the analyst cannot articulate how the method generated the specific results. Such methods thus cannot be used for hypothesis evaluation, description, or exploration (see Chapter 51), and they do not advance scientific understanding. Moreover, apart from measures of accuracy in classification or prediction, black box methods do not necessarily produce replicable substantive results (see Chapter 14).

Understanding is often elusive in science. In daily life, we rely on many established phenomena, such as an effective medicine or a physical law, even though scientists cannot account for the mechanisms involved. Nature may be difficult to understand, but it is impossible to understand when we use analytic approaches that are incapable of giving insight. Only transparent analytic methods are useful for scientific explanation.

56

Reasoning

... in the sciences the authority of thousands of opinions is not worth as much as one tiny spark of reason in an individual ....
—*Galileo Galilei, 1612*

The iron labor of conscious logical reasoning demands great perseverance and great caution ....
—*Hermann von Helmholz, 1862*

Fallacies ... do not cease to be fallacies because they become fashions.
—*G. K. Chesterton, 1930*

... you mustn't cheat by inventing as many hypotheses and approximations as you have experimental facts.
—*Philip Anderson, 1983*

Scientific reasoning may be innate in humans. Children use scientific practices and principles naturally, without instruction, as they learn about the world. Likewise, for many millennia, people have relied on scientific reasoning to survive, such as when tracking animals in hunting.

Yet we humans are also naturally susceptible to many logical fallacies that interfere with scientific reasoning. We are a social species. We depend on each other for survival. We are hierarchical. Even in the simplest societies or in small groups, we listen to and follow some persons more than others, sometimes voluntarily and sometimes not. Deception is common in our relationships with each other. Traditions and customs pervade our lives and represent the beliefs not only of the living but of our ancestors as well. We conform our thoughts, usually unconsciously, to those of our fellows, making our social relationships strong and effective. Due to these aspects of human nature, we easily suffer collective illusions.

Many formal and informal logical fallacies exist, and they dominate most human discourse outside of science. Not only do others' irrational thoughts surround us, we scientists *also* often think irrationally in daily life.

No matter how technically sophisticated our research is, bad reasoning can undermine it, and even make it worthless. To think logically, we must consciously train ourselves and work hard to overcome our vulnerabilities. I recommend reviewing an extensive list of fallacies (see Notes) regularly to stay on alert for them in your own and others' work.

General fallacies

A few general fallacies are common in scientific debates (see Chapter 43). The argument from ignorance often figures prominently. This argument is a claim that a proposition is true if it has not been proven false, or vice versa. For example, researchers may assert the presence of a hypothesized phenomenon because there is no evidence invalidating it, despite a lack of research focused directly on the phenomenon. Or researchers may dismiss a hypothesized phenomenon because no evidence of it exists, even though scientists have not sought evidence of the phenomenon through rigorous research. Without research on a hypothesized phenomenon, there is no rational basis to claim that it does or does not exist. In this circumstance, the maxim "absence of evidence is not evidence of absence" holds. However, it *is* very rational to question the existence of the hypothesized phenomenon when researchers do study the hypothesized phenomenon directly and rigorously, and consistently do not find evidence of it.

The appeal to the majority is a claim that a proposition is true if many people believe it is. Scientists make this appeal when they assert truth on the basis of, for instance, so-called consensus statements (see Chapter 15), seeming agreement among nearly all researchers in a field, or features in common among most theories on a topic. Conversely, researchers may irrationally declare an idea false because it is unpopular.

The argument from authority is a claim that a proposition is true because of the high positions of those asserting it. Scientists use this argument when they assert truth on the basis of claims made by eminent researchers or published in prestigious journals. This argument also applies when scientists assert truth on the basis of statements by professional societies, governments, and international agencies. In addition, scientists appeal to authority when they dispute claims of truth because low status researchers made them or low status journals published them. Whether explicit or implicit in researchers' work, these fallacies inhibit scientific progress because they have nothing to do with the substance of ideas and evidence.

Causal fallacies

We scientists frequently succumb to fallacies when thinking about causal relationships. In scientific research, particular fallacies often correspond to attempts to infer causality illogically from particular kinds of research designs. We can infer causality *only* from evidence in controlled experiments (see Chapter 52). Thus, reasoning about observational research in relation to causal hypotheses is more likely to be fallacious than reasoning about experimental research. At one extreme, researchers may analyze a case study, case series, or characteristics of some phenomenon and note a simple co-occurrence: characteristic X appears in case or phenomenon Y. From this result, they may claim X causes Y. Co-occurrence is not even correlation, much less causation. To establish correlation, researchers also need data on characteristics when the phenomenon is not present, such as in non-cases.

Of course, correlation in observational research also is not causation. Suppose variables A and B are correlated. Although A *might* cause B, there are many other possible explanations for their relationship. Perhaps another variable, C, causes both A and B, thus producing the correlation between A and B. Or maybe B causes A instead of A causing B. In another explanation, A might cause C and C might cause B, making A an indirect cause of B. A further possibility is that A causes B sometimes, depending on the presence or level of C. Either of these last two explanations may be possible even with experimental evidence that A causes B, if the researchers did not also control C in their experiment.

Researchers sometimes label results that are the inverse of an hypothesized causal relationship as paradoxical or ironic effects. Such patterns, if genuine, could imply a contingent causal relationship (A causes B depending on the presence of C), or a weak or even a non-existent causal relationship. Merely labeling the pattern, though, is no explanation of it.

In observational research, scientists often interpret longitudinal associations between A (measured at one time point) and B (measured at a later time point), as implying A causes B. This is wrong because without experimental control, it is impossible to know whether unmeasured variables could account for the longitudinal correlation between A and B.

Different hypothesized causes of an effect are not necessarily in conflict. There may be multiple causes of an effect. Some may act directly and others indirectly. Or they may operate on different levels of analysis or different scales of time or space. Likewise, a cause may have multiple effects.

In most fields, scientists are very familiar with these causal fallacies. Despite this, we fall prey to them often. To overcome them, it requires conscious effort. Whenever you propose or assess a causal interpretation, check whether the explanation suffers from any of the causal fallacies and whether alternate explanations are equally valid (review this chapter if necessary). Also, whenever you consider causal processes involving more than two factors, diagram the hypothesized causes, effects, and contingencies. The diagram will make it easier to understand the explanation and find logical errors in it.

Other irrationalities

Beware of other types of reasoning errors linked to specific scientific practices and principles. For example, to be useful in science, hypotheses must be falsifiable and come *before* making or analyzing observations (see Chapter 51). In some fields, many researchers neglect these truisms, deluding themselves that ideas violating these precepts still count as hypotheses.

In attempting to explain a phenomenon, always try to account for it first with known phenomena before invoking some new hypothesized mechanism, process, element, or force. In their zeal to promote a new hypothesis, scientists sometimes neglect previously studied factors, even mundane ones, including artifacts, as explanations. Search the literature thoroughly (see Chapter 5) before proposing a new phenomenon.

Watch out for many other kinds of reasoning errors common in scientific research:
- selecting some evidence but ignoring other evidence (see Chapter 26);
- making claims beyond the limits of generalization (see Chapter 21);
- mischaracterizing how (in)consistent results are across studies (see Chapter 15);
- making claims about the empirical world based on little or no empirical evidence (see Chapter 54);
- failing to account for the broader context of a problem or the full range of possible variation (see Chapter 57); and
- blaming your data, method, or design for unexpected results (see Chapter 21).

It is easier to see irrationality in others than in ourselves. Therefore, seek critical feedback on your work (see Chapter 31) to root out your reasoning errors. Although we can point out others' apparent departure from reason, we have no control over them. When the scientific or other stakes are high, it is likely that irrationality will persist, especially if it serves to protect dominant points of view or other interests. Strive for reason in your own work, as that is the only work you can control.

57

Considering the Unseen

Stop there! your theory is confined to that which is seen; it takes no account of that which is not seen.
—Frédéric Bastiat, 1850

The great difficulty is in trying to imagine something that you have never seen, that is consistent in every detail with what has already been seen, and that is different from what has been thought of ....
—Richard Feynman, 1963

As humans and scientists, we often blind ourselves to causal processes by narrowing the scope of a problem artificially. We notice the effects, but do not see the causes. Constants—factors always present or always absent—in our focus of study often lead us to ignore such factors as *variables*, which can change and thus potentially influence the outcomes we observe. That is, we might not recognize contingencies because some preconditions are not present in our limited field of view. We may accept past events or current circumstances as givens, forgetting that they are not universal conditions and that had they been different the outcomes might also have been different. We may even come to believe that the constants we observe are, in principle, truly constant because they cannot be manipulated in practice due to political, ethical, technical, or other constraints.

The unseen also includes the full cascade of consequences from a cause. We are most likely to recognize and postulate about immediate effects. We tend not to contemplate delayed effects, the effects of the effects, and further downstream impacts even though they may be just as observable. Similarly, we often trap ourselves by framing a causal chain as a directional sequence with a beginning and an end. With this approach, we neglect possible antecedents to the beginning cause, and may miss loops and cycles in the full process.

Furthermore, interactive effects may escape our view without our focused attention and thought. That is, causal relationships might only appear under certain conditions, or the relationships might change depending on those conditions.

Researchers also disregard the unseen when they forecast the consequences of change in natural, social, and other complex systems as being overwhelmingly positive or negative. Such lopsided outlooks typically reflect a narrow, short-range focus or ideological bias.

As Feynman noted, considering the unseen is perhaps one of the biggest intellectual challenges in science. It involves conducting thought experiments spurred by the question "what if?" in relation to factors and relationships outside the bounds of prior research or assumption. Such exercises can feel like fantasy, yet they are also responsible for many scientific insights.

Another way to consider the unseen is to conceive of a research topic as broadly as possible and imagine extremes of comparison. By studying a topic across the spectra of time, space, and kinds (e.g., species, cultures, environments, etc.), we may observe increased variation, which gives more contrast to what had once been unseen and allows us to detect it more easily.

58

Judging Hypotheses

Not even observation itself, which by its very nature is passive, can be realized without hypotheses. Good or bad, a conjecture (or any attempt whatsoever at explanation) should always be our guide. No one searches without a plan.
—*Santiago Ramon y Cajal, 1916*

Facts do not 'speak for themselves.' They speak for or against competing theories … facts force us to discard some theories — or else torture our minds trying to reconcile the irreconcilable ….
—*Thomas Sowell, 1987*

… no good model ever accounted for all the facts, since some data was bound to be misleading if not plain wrong. A theory that did fit all the data would have been 'carpentered' to do so and would thus be open to suspicion.
—*Francis Crick, 1988, quoting James Watson*

We scientists are constantly judging hypotheses—considering them and forming opinions about them. This mental activity is distinct from hypothesis evaluation, in which we conduct studies designed to test hypotheses (see Chapter 51). Of course, such research is critical for judging hypotheses, but judging also involves considering descriptive research, reasoning (see Chapter 56), and other abstract matters.

We judge hypotheses in many different contexts, such as when casually scanning empirical results in the literature, preparing introduction and discussion sections of research reports, and debating with other scientists (see Chapter 43). The key elements of judging hypotheses are the same as those for judging theories and models.

Multiple objective criteria are relevant for judging hypotheses. Yet judgments of hypotheses are subjective, as scientists vary in how they interpret, weight, and apply the criteria.

Consistency with evidence

...truth is correspondence with the facts.
—*Karl Popper, 1959*

A primary criterion is the degree to which an hypothesis is consistent with the empirical evidence. However, not all evidence is equal.

Scientists develop hypotheses initially to describe or explain prior observations. When an hypothesis fits with those observations, scientists often keep the hypothesis and continue to compare it informally with other existing evidence. Usually, when a developing hypothesis clashes with prior results, scientists discard or change it privately. Thus, researchers tend to shape their hypotheses in response to the existing evidence before proposing them publicly. Exploration (see Chapter 51) and model building in simulation (see Chapter 54) are also examples of this process. Empirical observations that researchers use to develop an hypothesis, then, cannot be used meaningfully at any later point in judging the hypothesis' consistency with evidence. We can judge an hypothesis only by comparing it with *new* observations.

It is pointless to judge an hypothesis without considering all relevant evidence. Ensure that you assemble the evidence comprehensively (see Chapters 5, 15, and 26) before forming a strong opinion on an hypothesis.

Poor quality evidence can make it difficult to judge an hypothesis accurately. Given scientists' attachments to particular hypotheses (see Chapter 53), the published empirical record may be biased in favor of or against an hypothesis. The value of the consistency between an hypothesis and evidence is proportionate to the amount of evidence based on pre-registered (see Chapter 4) and independently replicated (see Chapter 14) research involving appropriate data analysis (see Chapter 8) and publicly available data (see Chapter 9).

The importance of consistency with evidence also depends on the rigor and relevance of the designs and methods in the underlying research. For instance, experimental research gives a strong basis for assessing the consistency of a causal hypothesis with evidence, while observational research does not (see Chapter 52). In another example, errorful measurement can obscure or produce consistency between an hypothesis and evidence.

Other aspects of empirical evidence pertain to judgments of hypotheses. Studies based on different methods, designs, and contexts comprise multiple lines of evidence that strengthen empirical assessment of an hypothesis. When there are rival hypotheses, head-to-head evaluations of them within studies may provide especially critical evidence on the hypotheses' relative merits.

Constancy

When the results of the first experiment or observation fail to support the hypothesis, instead of abandoning it altogether, sometimes the contrary facts are fitted in by a subsidiary clarifying hypothesis. This process of modification may go on till the main hypothesis becomes ridiculously over burdened with ad hoc additions. The point at which this stage is reached is largely a matter of personal judgment or taste.
—W. I. B. Beveridge, 1950

Researchers often revise their hypotheses in response to inconsistency with new observations. Initially, small revisions may be reasonable, as it may reflect researchers' improved understanding of the problem or correction of an error in thought or expression. However, if researchers revise substantially or continue to revise, the hypothesis effectively becomes a new hypothesis. This means that prior versions of the hypothesis are not viable and may imply that the current version is fundamentally weak, too. A rational approach is to accept the demise of the hypothesis and develop a different idea. An hypothesis that changes little over time therefore indicates it is relatively free of opportunistic tampering by researchers.

Generality

An hypothesis that applies broadly or universally to its corresponding problem makes multiple, narrower hypotheses unnecessary. Similarly, including many specific conditions for an hypothesis may imply that it does not describe a genuine phenomenon or misses important aspects of the phenomenon.

Arbitrary elements

Models often have arbitrarily defined elements, such as constants, that researchers include to improve model fits to data. Deriving elements from intuition, deduction, and/or empirical observation is appropriate, but assigning them merely on mathematical or statistical grounds simply signals the model is incomplete at best or misleading at worst.

Parsimony

> ... in general, we consider it a good principle to explain the phenomena by the simplest hypotheses possible ....
> —Ptolemy, c. A.D. 100–170

Hypotheses that have few elements or connections among elements are parsimonious. With all else equal, simple hypotheses are easier to communicate, think about, and evaluate than complex hypotheses (see Chapter 13).

Irrelevant criteria

In judging hypotheses, many researchers invoke subjective criteria, such as plausibility, potential to explain other phenomena, or fit with predominant hypotheses and theories. These criteria simply cloak researchers' preferences and prejudices, and serve as excuses to stop considering a threatening hypothesis or accept a comforting one with little scrutiny. Scientists also often lapse into formal and informal logical fallacies (see Chapter 56) that poison judgments of hypotheses.

Strategies for judging effectively

If you do not expect the unexpected, you will not find it ....
—Heraclitus, 535–475 B.C.

I have steadily endeavored to keep my mind free so as to give up any hypothesis, however much beloved (and I cannot resist forming one on every subject), as soon as facts are shown to be opposed to it.
—Charles Darwin, 1881

I don't see the logic of rejecting data just because they seem incredible.
—Fred Hoyle, 1976

The interplay between hypothesis and evidence is continual. When we encounter new observations on a particular problem that are inconsistent with or beyond the scope of an hypothesis, it is easy to ignore or dismiss them. Instead, note that the hypothesis is inconsistent with or does not extend to the observations. But don't stop there. Consider other hypotheses or develop new ones, and judge them in relation to the new observations. To understand nature, we must be open to all hypotheses and seek to conjure them often. Without this mental flexibility, we will not be able to see evidence relevant to the problem. I know I am considering ideas without restriction when I try on ideas that violate my own preferences, beliefs, and inclinations.

When you judge hypotheses in a planned way, literally write them out, and note evidence consistent and inconsistent with each. It is too easy to discount and overlook evidence when judging hypotheses mentally only. Compare the evidence for rival hypotheses on a side-by-side basis with the same measures and criteria.

It is impossible to judge an hypothesis fairly without considering all of the relevant evidence. Search for and summarize evidence systematically and comprehensively (see Chapters 5 and 15).

When judging an hypothesis, assess the quality of the evidence, not just its consistency with the hypothesis (see Chapter 52). Before you compile the evidence, state explicitly the relative strength of different possible kinds of evidence, no matter whether any of these kinds are available. This puts your judgment in context, and leads you to highlight limitations and calibrate your confidence in your overall interpretation. Keep in mind that the quality of evidence extends beyond methodological and design factors and includes the other essential practices and principles covered in this book. Use a checklist (see Chapter 17) to ensure you cover all pertinent aspects of quality.

Review descriptions of logical fallacies regularly (see Chapter 56) to recognize and avoid them, as logical errors are common when judging hypotheses. Also beware of other pitfalls. An hypothesis cannot dismiss or invalidate another hypothesis; only empirical evidence can nullify hypotheses. Likewise, an hypothesis that invokes hypothesized phenomena as causes is actually multiple hypotheses. If only one part of the hypothetical proposition can be evaluated empirically, no judgment can be made on the other part that has not been or cannot be evaluated empirically. Furthermore, be alert to anomalies, or results unexpected by one or more hypotheses.

Resist the temptation of concluding that anomalous results simply reflect error or chance variation. They well might, but if the anomalies appear repeatedly, investigate thoroughly.

Beware of your own philosophical or personality dispositions that can interfere with rational assessment of hypotheses. Many people have a tendency toward moderation and dislike extremes. Such an inclination can lead to seeking a middle ground between competing hypotheses. A similar orientation involves believing that all or many ideas have value, which can lead to concluding that multiple, even contradictory, hypotheses are useful. Still another tendency is a fondness for extreme views. This can lead to favoring an hypothesis simply because it is bold, sweeping, or unpopular. Identify your own disposition and recheck your assessment of hypotheses on objective criteria whenever your conclusion matches that disposition (see Chapter 61).

Hope for and wishful thinking about particular hypotheses can also undermine our judgments of them. These emotions are strongest when an hypothesis has significant practical consequences, such as for a disease, ecological problem, or economic growth, even apart from the influence of our ideological commitments or self-interest. We may think an hypothesis must be true because otherwise catastrophe awaits or the chance for progress of some sort vanishes. To judge hypotheses rationally, find a way to neutralize, or least isolate, such fears and wishes.

As scientists do more research on a problem, evidence that once was equivocal may become more decisive, one way or another. Reconsider an hypothesis as more evidence related to it accumulates, regardless of your prior opinion of the hypothesis. Also, recognize that sometimes there are no good existing hypotheses for a problem. Don't just settle the best of a bad lot. Instead, develop new hypotheses and evaluate them empirically.

59

Curbing Scientific Chauvinism

How many modern scientists have convinced themselves that they are now the heirs of creation, that society owes them a living and must provide them with the facilities they need to further their researches? ... In fact the world owes a living to no one who does not earn it, and whose efforts have no economic significance. Intellectual achievements may be rewarding to the individuals, and vitally important to society, but they must be judged on their merits; no man must be the sole judge of his own cause.
—*Lord Bowden (Bertram Vivian Bowden), 1965*

... we must constantly remind ourselves (especially in connection with the application of science) of the finitude and fallibility of our knowledge, and of the infinity of our ignorance.
—*Karl Popper, 1971*

Many scientists have outsized opinions of the importance of science and scientists. Such opinions can be unproductive and divert scientists from doing good research.

As Terence Kealey showed in *The Economic Laws of Scientific Research*, most major practical advances in technology come from tinkering and informal trial and error—not scientific research and theoretical understanding. Throughout history, technological advances tended to enable scientific progress. Of course, there are some genuine technological breakthroughs originating from scientific research. These, however, are disproportionately serendipitous, not the products of planned research programs. Applied research can produce incremental improvements to technology and practice, but typically not revolutionary transformations. Technological advances come primarily from efforts to develop technology directly, not from scientific research (at least not from what scientists consider scientific research).

The ultimate value of science is understanding. Practical applications are an occasional windfall.

Scientists (and their institutions and funders) frequently oversell the potential practical applications of their work. It is a common tactic for getting funding and attracting attention, but usually amounts to deceptive marketing. A brief review of history shows that such projections typically are very over-optimistic and rarely fulfilled even partially.

Greater humility about what science actually contributes might serve to give non-scientists a better view of science and instill greater freedom of thought among scientists. Humans made astounding technological and other advances for many millennia before formal science even came into being. Human ingenuity, not science, has been the driver of progress.

How many scientists, especially those in academia, can say their work has directly led to practical applications that are implemented in the real world, regardless of impact or ethics? That is, contributions that were not just applicable, but actually applied on more than a temporary basis? (Patents, press coverage, and people [politicians, bureaucrats, or others] talking about specific research are not examples of actual application.) I estimate that those who can answer these questions affirmatively make up a very small fraction of scientists, including those exclusively involved in applied research. There are whole *fields* in which research has not led to any meaningful practical applications. And in many, perhaps most, cases where non-scientists have applied scientists' work enthusiastically, scientists designed the research specifically to produce such incremental applications.

Governments now fund vast amounts of applicable and applied research, much of it relevant to government actions and policies. Yet in most cases, research relevant to government never gets applied due to political, bureaucratic, and other factors. Such applied research generally goes unapplied because it was not developed in response to true demand (especially for government-funded research for potential commercial applications). Although we scientists can do applied research on questions that interest us, often there is no one in the real world (or government) who is able and invested enough in the topic to apply the results. Scientists' ideas for potential applications also tend to be uneconomic or otherwise impractical. Applied research in industry or as part of entrepreneurial ventures is much more likely to be applied precisely because there is a clear commitment and ability to implement the results. Industrial firms also usually assess the viability of potential applications before beginning research related to them.

Even in the strictly scientific realm, new scientific developments tend not to be taken up by others. In most fields, most new methods are not used or are used rarely by other scientists. New empirical results also tend not to be pivotal discoveries, and most studies are not replicated. New hypotheses often are never evaluated.

The value of new scientific developments is general: each adds to the warehouse of knowledge and has the *potential* to be applied in scientific or practical contexts. Sometimes the application may occur long after the research was done. Regardless, application of research is not the norm, and is not something scientists usually can influence. Consequently, it is helpful to have reasons other than application for doing research (see Chapter 60).

Many scientists and their supporters advocate for government science funding as being essential for economic vitality. The historical record, as reviewed by Kealey, shows that this is also not so. By perpetuating the myth that science leads to economic growth, scientists inflate their relevance and foster their own arrogance, ultimately harming the reputation of science and scientists as a whole. Moreover, Kealey found that government funding may actually hamper scientific development.

It is easy to fall into the mental trap of expecting *someone*—governments, foundations, corporations, philanthropists—to fund scientific research because it is so important (especially when it is our own work!). I suffered from this delusion much of my career. However, my mental health improved when I lost my sense of entitlement and focused my energies on doing research, funded or not (see Chapter 16).

60

Getting Immediate Rewards from Doing Research

... in an absolute sense, science is good in itself, apart from its [content of] knowledge; ... its lure is everlasting and unbroken ... [the servant of science] should praise the assiduous whenever their effort is for delight [in science itself] rather than from [the hope of achieving] victory in argument.
—Abu Raihan al-Biruni, A.D. 973–1048

... my work, which I have done for many a long year, was not pursued in order to gain the praise I now enjoy, but chiefly from a craving after knowledge ... whenever I found out anything remarkable, I have thought it my duty to put down my discovery on paper, so that all ingenious people might be informed thereof.
—Antonie van Leeuwenhoek, 1716

We must not be discouraged if the products of our labor are not read or even known to exist. The joy of research must be found in doing since every other harvest is uncertain ....
—Theobald Smith, 1933

What's the use of doing all this work if we don't get some fun out of this?
—Rosalind Franklin, 1920–1958

To stay happy as a scientist, it is essential to get immediate rewards from doing research. By immediate rewards, I mean those you receive while doing the research work itself. Some examples of immediate rewards are enjoyment in doing specific research tasks and fun in learning something new from the research. Pleasure from working with your colleagues and conducting research as a favor for someone also count as immediate rewards. Direct compensation in the form of money or new skills are kinds of immediate rewards, but for many scientists, money alone is not enough to stay happy in research over the long term.

Sometimes, researchers are not free to study what they want or are restricted (for example, by contract, budget, or management decree) how they conduct their research. Such limitations naturally affect many of the immediate rewards a scientist might receive from doing research. Not all contexts for doing research are the same, making it important to choose your work settings carefully.

Immediate rewards are the only ones you can be fairly certain you'll get, as Smith emphasized. Other potential rewards that might come in the future, however, often are unrealized, illusory, or unlikely to satisfy.

Scientists tend to have lower salaries than many other professionals. Recognition in the form of special awards, invited keynote and other important lectures, fancy job titles, and election to elite scientific groups substitute, in some sense, for money as incentives to researchers. But these kinds of rewards and other more informal recognition are scarce. As a result, most researchers will never receive them. Such recognition and prestige also come, if ever, years or decades after the work that supposedly underlies them.

Many researchers consider the number of citations a scientific work receives as an indicator of its quality and importance. By extension, they often regard the number of citations a researcher receives overall as reflecting his or her scientific status. Yet scientists often cite sources mindlessly, without even reading them (see Chapter 26). Many of my colleagues and I have noticed that perhaps half or more of the citations our publications (in diverse fields) receive are erroneous. Sometimes authors cite us for a finding or point *opposite* to what we reported. Other times, authors attribute to us something entirely unrelated to our work. Moreover, plenty of accurate citations are trivial, equivalent to "so and so studied X," without referring further to the content in the cited source. It is impossible to interpret the significance of a citation without taking context and validity into account. Counting citations is a hollow exercise, especially for the researcher looking for genuine rewards.

Applications of scientific work are even rarer than citations to it. Many researchers crave to be known for starting fields or blazing new trails of investigation that others follow. Usually, many scientists together, in uncoordinated or loosely coordinated fashion, create the new fields and trails, with the contributions of any particular individual difficult to identify objectively. New fields sprout very infrequently and most new trails of investigation attract no followers. The impacts we researchers have on our fields typically are quite modest and hard to discern with confidence.

Application of scientific research in the real world is still more uncommon. Outside of industrial or entrepreneurial ventures, applied scientific research is rarely applied in the real world (see Chapter 59). Even though governments fund much applied and applicable research, political and bureaucratic forces work against actual application (see Notes). Apart from such barriers, scientists' ideas for application also tend to be impractical or uneconomic, perhaps because we lack sufficient exposure to real world contexts and constraints (see Chapter 2).

In short, how others use or interpret your work is almost entirely out of your control. Disappointment looms for scientists who seek rewards based on reactions to their research. Fulfillment in science rests on having an attitude that Homi Bhabha expressed: "My success will not depend on what A or B thinks of me. My success will be what I make of my work."

How can you know whether the immediate rewards from doing research are enough for you? Ask yourself: if none of your dreams or goals about what will happen *after* your research is done were ever achieved, would it still be worth it to you to do the research? That is, would you still do the research if, after you complete and report the research, no one else *ever* read your reports or listened to your presentations, much less cited or acted on them? Happy scientists answer "yes."

And why wouldn't we? The pursuit of truth is thrilling in itself, as Fridtjof Nansen described so beautifully in 1897: "… we throw ourselves heart and soul into these very researches, consumed with a burning thirst, to absorb everything into ourselves, longing to spy out fresh paths, and fretting impatiently at our inability to solve the problem fully and completely."

61

Doubting and Questioning

Constant and frequent questioning is the first key to wisdom ... For through doubting we are led to inquire, and by inquiry we perceive the truth.
—*Peter Abelard, 1079–1142*

The first condition to be fulfilled by men of science, applying themselves to the investigation of natural phenomena, is to maintain absolute freedom of mind, based on philosophic doubt ... When we propound a general theory in our sciences, we are sure only that, literally speaking, all such theories are false. They are only partial and provisional truths which are necessary to us, as steps on which we rest, so as to go on with investigation; they embody only the present state of our knowledge, and consequently they must change with the growth of science ....
—*Claude Bernard, 1865*

... every new fact must come under merciless scrutiny, every step in reasoning under meticulous criticism. Only those who have shared in this activity can understand the joy of it. Science is a living thing, not a dead dogma.
—*Cecilia Payne-Gaposchkin, 1979*

If you don't really stop and think about things, it doesn't matter whether you're a genius or a moron.
—*Thomas Sowell, 2013*

Doubt is the lifeblood of science. Without doubt, we would never question observations or ideas. Progress of any kind would be impossible.

If scientific knowledge does advance, then it must change. If most or all current knowledge must change (if it is ever to advance), then it means most or all current and future knowledge is *wrong* or *incomplete* in some major way. Therefore, the most rational stance as a scientist is to doubt all or almost all explanations, interpretations, and evidence in science, as they are all tentative.

Although limited data can constrain scientific progress on a problem (see Chapter 48), limited questioning by scientists may harm the pursuit of truth more. Decades, centuries, or millenia can pass with effectively no new crucial data on a problem. Yet once scientists begin questioning old views of those data, their understanding of the problem can improve dramatically and rapidly.

In science, doubting and questioning are nearly inseparable. Doubting involves acknowledging uncertainty and distrusting the accuracy and completeness of ideas and evidence. This leads, almost automatically, to searching for and generating alternate possibilities and explanations in the form of questions. Scientific doubt isn't *resistance to* different ideas—it is the *consideration of* different ideas. Disbelief, dismissal, and disregard are the opposites of scientific doubt. They involve no questioning or evaluation of other ideas. Rather, they are anti-scientific thoughts and actions that scientists and others use to quell doubt and doubters.

Doubt applies to *all* ideas and evidence—new and old, our own and others', and those offered by majorities and minorities of researchers. The time for doubting and questioning on any scientific topic is never past. There cannot be too much doubt or too many questions. No *scientific* harm ever comes from doubting and questioning; to the contrary, their lack kills science.

Even if every scientist doubted and questioned consistently, there is no risk of science becoming a chaotic mess with researchers incapable of agreeing on anything. We can develop ideas and test them with empirical evidence while doubting all along. Honest, critical scientists recognize regular patterns in evidence, and their skepticism makes their converging opinions more powerful. Furthermore, our human tendencies to trust and find common ground with each other are much stronger than our impulses to doubt, no matter how much we try to strengthen them.

Highly intelligent persons may be more likely to engage in deception and self-deception than others. To the extent scientists have above average intelligence, we may be especially prone to these traps. Doubting and questioning are the only antidote to the psychological and sociological demons that thwart our pursuit of truth.

Doubting yourself

The greatest deception men suffer is from their own opinions.
—*Leonardo da Vinci, 1452–1519*

Let us subject ourselves to harsh self-criticism that is based on a distrust of ourselves. During the course of proof, we must be just as diligent in seeking data contrary to our hypothesis as we are in ferreting out data that may support it. Let us avoid excessive attachment to our own ideas, which we need to treat as prosecutor, not defense attorney ... It is far better to correct ourselves than to endure correction by others ... We must admit our absurdities whenever someone points them out, and we should act accordingly
....
—*Santiago Ramon y Cajal, 1916*

The first principle is that you must not fool yourself, and you are the easiest person to fool ... After you've not fooled yourself, it's easy not to fool other scientists.
—*Richard Feynman, 1974*

You are the first line of defense against producing nonsense. You are the only critic that you can guarantee will doubt your ideas and evidence. After publishing or presenting your work, other researchers may not doubt your ideas and evidence sufficiently or well. If you can make self-doubt a habit, then you can also extend your doubt easily to others' ideas and evidence.

Strategies

Remind yourself that all hypotheses and theories are provisional and almost sure to be wrong, at least partially. Your ideas are only means for pursuing truth, but not ends in themselves. This mindset frees you psychologically. By accepting that your ideas are wrong, you can focus on how they are wrong, which also makes criticism from others easier to bear.

Don't rush your work, as self-doubt is often the first casualty of working too fast. In scientific research, quality always trumps quantity. Researchers who seek to publish many papers for the sake of having many publications are their own worst scientific enemies. It is far better to complete your work when you have had the time to thoroughly question it. Indeed, this is why Charles Darwin took 20 years to publish *On the Origin of Species*.

Systematically review your work in all stages of research, searching for errors in research design (see Chapter 3), data (see Chapters 6 and 7), programming code (see Chapter 11), analysis (see Chapter 8), interpretation, and assumptions (the last particularly for those underlying your research on a problem). Unless you do so consciously, it is likely you will miss problems and underestimate your own fallibility.

As you draft scientific reports and other documents, question each sentence. Is it *entirely* accurate? Is there *any* uncertainty to the point? If so, revise accordingly. Non-scientists often criticize scientists for their obsession with precise details and tentative phrasing. These traits are valuable because they reflect researchers' doubt, which drives scientific progress.

Often revisit matters on which you have previously questioned yourself, even for past research. It can be difficult to admit when we are wrong, and our egos protect our closely-held ideas strongly. Our irrational defenses are more likely to fall after repeated attacks of doubt. In my experience, the threatening feeling of possibly having made a mistake becomes less with each episode of self-questioning. I try to take it as a challenge to defeat myself, and find my errors. Any disappointment I feel about having made a mistake is always outweighed by the sense of relief that I have identified and corrected it.

Other strategies to increase self-doubt include avoiding research topics on which you have ideological conflicts of interest (see Chapter 48), detaching yourself from your hypotheses (see Chapter 53), and getting critical feedback on your ideas and work from others (see Chapter 31).

Researchers tend to defer to scientists in positions of authority and may be unlikely to challenge their ideas and evidence. High status scientists may feel, mistakenly, that they are less prone to error than others, when in fact they may be at greater risk because of this attitude and others' behavior toward them. To preserve doubting and questioning of your work, avoid high status positions. If you have a high status position, then redouble your efforts at self-doubt.

Indicators of self-doubt

How can you tell whether you doubt yourself sufficiently? One marker is discovering an error in your own work. Everyone makes errors. Even better than finding an error is discovering that what you thought was an error is not an error.

If your strongly held beliefs about the problems you study do not change on occasion (over years or decades) in response to your own doubts, others' criticism, or evidence, then you have not doubted meaningfully. If your strongly held beliefs have changed, but only as a result of majority opinion changing among those scientists with whom you interact or whose work you read, such shifts are not examples of doubt but of conformity. When doubt leads you to change your strongly held beliefs, you recognize it because it is the result of your conscious reasoning. The experience is usually painful or exciting, which etches the struggle into your memory.

Doubting others

... the duty of the man who investigates the writings of scientists, if learning the truth is his goal, is to make himself an enemy of all that he reads, and, applying his mind to the core and margins of its content, attack it from every side. He should also suspect himself as he performs his critical examination of it, so that he may avoid falling into either prejudice or leniency.
—Alhazen, c. 1025

... where men are the most sure and arrogant, they are commonly the most mistaken, and have there given reins to passion, without that proper deliberation and suspense, which can alone secure them from the grossest absurdities.
—David Hume, 1751

Oh comforting solitude, how favorable thou art to original thought!
— Santiago Ramon y Cajal, 1916

... you must doubt the experts.
—Richard Feynman, 1966

... the orthodoxy produced by intellectual fashions, specialization, and the appeal to authorities is the death of knowledge, and ... the growth of knowledge depends entirely on disagreement.
—Karl Popper, 1993

If you have assessed your own ideas and work and noticed errors, it is easy to imagine that others have similar problems. However, they might not have doubted themselves thoroughly, so as a consumer of their ideas and evidence, you must be responsible for doubting and questioning.

Seeking truth is an adversarial process. Karl Popper emphasized that critical discussion is the defining feature of science. Doubt fuels critical discussion, both within an individual (questioning your own ideas and evidence) and between individual scientists.

The popularity of an idea, method, or evidence does not indicate its merit or logic. It is prejudice, not doubt, to distrust a method, result, or idea based on the training (or lack thereof), status (or lack thereof), personality, demographic characteristics, reputation (or lack thereof), affiliation, academic lineage, or social circle of the person presenting it. It is also prejudice to question a method, result, or idea based on where it was presented or published.

Targets of doubt

All scientific work and ideas deserve doubt. However, scientists often do not doubt or question ideas and results with the following characteristics:

- work based on the dominant hypotheses, theories, and methods in a field (see next section on "Majorities and minorities");
- statements expressing researchers' high confidence (a symptom of both deception and self-deception), apart from technical descriptions of statistical probabilities; and
- work by high status researchers or organizations (to whom we often defer reflexively) and research reported in high status journals (with a likely greater chance of fraud and other misconduct than research in other publishing outlets; see Chapter 47).

Focus your doubt consciously on ideas, methods, and results with these characteristics to counter their subtle appeal.

Hypotheses, interpretations, and explanations are natural targets for questioning. Less obvious, but equally important, targets of doubt are definitions, assumptions, and calculations. Definitions of concepts, variables, measures, and values of variables can each change results and either create or eliminate phenomena for study by fiat. Assumptions can be difficult to question because researchers often do not articulate them completely or at all. As a doubter, consciously pick apart an idea, method, or result systematically by asking what definitions and assumptions are involved and critiquing each in turn.

Errors in data collection, management, and analysis are common, even for the most careful researcher. Consequently, any calculation has uncertainty on this basis alone and is worthy of questioning. Replication (see Chapter 14) and public archiving of data (see Chapter 9) ultimately may help detect such errors, but in the short term no particular result warrants high trust.

Many researchers also defer to the wizards in science, such as statisticians, mathematicians, computer modelers, and theorists, who use unfamiliar jargon, formulas, and techniques. These wizards are vulnerable to the same pressures as other researchers that can make their work biased or wrong. Just like other scientists, they can try to make insignificant matters significant when trying to draw attention to their work and prescriptions. As in all science, the ultimate proving ground is empirical evidence. Statistical, mathematical, methodological, or theoretical guidance lacking empirical and practical demonstrations of value often is useless. Muster your confidence and question wizardry just as you would other research.

Identify the assumptions and beliefs, often not articulated explicitly, that underlie research in a field or on a particular problem (see Chapter 48). Evaluate them critically. Too often we scientists accept the foundations of a field as received wisdom rather than as propositions to assess thoroughly and independently.

Strategies

When reading research reports or listening to presentations, presume the authors or presenters are wrong—misinterpreting or overstating—until they demonstrate otherwise with evidence and logic. This default orientation is especially important when assessing summaries of research in abstracts and media reports. This approach also lowers your cognitive burden. Rather than accepting some claim as true and then later rejecting it as not true, you can set a high bar in your mind for what you might consider provisionally true. Whenever I listen to a scientific presentation or read a research report, I question and argue with the presenter or author in my mind. Doubt is the intellectual immune system—it helps prevent infection with bad ideas, weak evidence, and illogic.

Similarly, assume that the authors are committed to their hypotheses or interpretations of the results, no matter what the results might show. Many, if not most, scientists approach their research in this way (see Chapter 53). Consider research reports as sales pitches in which the authors try to persuade you of their messages. Your task is to evaluate their claims against the evidence they show, while suspecting their biases throughout.

Consider ideas and evidence one piece at a time. Don't rush ahead to evaluate the conclusion. By taking each part of an idea or evidence in turn, you narrow the scope and make the task of evaluating (and doubting) easier. If you jump to the conclusion —your own or others'—you are more likely to accept or reject the idea or evidence wholesale, often on the basis of the implications alone. The implications may relate to your theoretical biases, non-scientific beliefs, economic and professional self-interest, and/or attachment to results from your past work.

Doubting and questioning are not just negative, critical activities, in the sense of tearing apart evidence or an idea. If you are skeptical of *all* ideas, then *all* ideas are worth considering. Examine each contending idea with an open, but critical, mind. This also means that when you think about a new hypothesis or interpretation of evidence, generate the whole idea before you begin evaluating it (see Chapter 45). Most of our ideas are wrong or bad in some way. Experiencing this low rate of good ideas reminds us that we are often mistaken and need to doubt our own and others' ideas vigorously and continuously.

Extensive and intensive doubting usually requires solitude. Because of social influence and dominance relations, we tend to doubt and question the least in group settings. Conferences tend to be echo chambers even when people with diverse views gather. Many, perhaps most, scientists are uncomfortable challenging others, even just mentally, in group contexts. It is easier to think critically of others' work when not in their presence. Frequent conferences and meetings of all types not only reduce the time for actual research, but also interfere with doubt, with significant harm to scientific progress.

When doubting others, we also get a chance to doubt ourselves. Both in science and our personal lives, we often criticize others for what we do ourselves. One way to lessen such hypocrisy is to turn every criticism you make of others on yourself. Do you suffer from the same flawed reasoning you have seen in another? Have you also ignored the assumptions underlying your research? Do you use the sound research practices you urge others to use? The solution to any discrepancy is to change your behavior to be consistent, not avoid questioning others. Doubting others thus can be an excellent way to boost and discipline doubting yourself.

Avoid group identities

Your most important identity in scientific research is as an *individual* scientist. Avoid identifying, at least consciously, with any *group*, such as a discipline or specialty of researchers, institution, professional society, committee, or country. Similarly, avoid identifying with demographic categories, such as by sex, ethnicity, or generation. Groups, and group psychology, have been central in human evolution. They have had and continue to have both positive and negative effects on people. But in science, groups usually are harmful to pursuing truth.

Our group psychology leads us to divide our fellow human beings into "us" and "them." Scientific research, however, is not about persons or groups, but ideas and evidence, no matter who presents them. If we define or perceive ourselves, and thus others, as members of key groups, we can no longer judge ideas and evidence rationally. Instead, we favor the ideas and evidence from our own group, and disfavor those from other groups. We tend to see the creativity and impact of research done by members of our group, and tend to dismiss the research by members of other groups as an undifferentiated mass of bad or irrelevant work. That is, we do not doubt the work in our own group enough, and we don't consider the work in other groups enough even to begin doubting it specifically. Pride in group membership is poison in science. Celebrate logic and sound scientific practices instead.

In science, especially on contentious topics, groups often do develop around different ideas or approaches. Those who share an idea or approach then interact and communicate more with each other than do those who have different ideas or approaches. Depending on the context, such groups can help or hurt the pursuit of truth (see Chapter 62). In principle, these groups correspond to scientific differences of opinion. In practice, though, the groups often are ultimately based on non-scientific concerns, such as funding and other economic interests. Regardless, even these groups damage the search for truth to the extent that the group dynamics inhibit communication and consideration between opposing sides.

Majorities and minorities

Whenever you find that you are on the side of the majority, it is time to reform—(or pause and reflect).
—Mark Twain, 1835–1910

Nature does not respect democratic methods or opinions.
—Donald Braben, 1994

Intolerant dogmatism, however, is one of the main obstacles to science. Indeed, we should not only keep alternative theories alive by discussing them, but we should systematically look for new alternatives. And we should be worried whenever there are no alternatives — whenever a dominant theory becomes too exclusive. The danger to progress in science is much increased if the theory in question obtains something like a monopoly.
—Karl Popper, 1994

Majorities in both thought and practice are dangerous to all scientists. As social animals, we are especially attuned to coordinating our thoughts and behavior with others, commonly through imitation. We are vulnerable to conforming to others on matters big and small, regardless of whether social pressure is real, imagined, or absent entirely. Even emerging majorities, such as scientific fads, exert the same kinds of forces as long-standing majorities.

It is comfortable to be in the majority. Psychologically, majorities often are like dominant groups. When we are in the majority, we are prone to viewing those in minorities as belonging to inferior and foreign groups. We may regard minority ideas and results to be of so little importance that we do not consider them enough to doubt them, and instead ignore or dismiss them outright.

Similarly, when we are in the majority, we are unlikely to doubt the majority beliefs and results. They are mostly safe from attack as we encounter few skeptics. Others' support of the majority leads us to be less critical of it. We seem to tell ourselves that if so many others hold an opinion or use a method, then it must be reasonable and sound.

Majority theories and opinions among researchers become default features of the intellectual landscape on a problem with many researchers ceasing to differentiate them from fact. Apart from group psychology, holding a majority opinion can make it difficult to entertain minority opinions on the same topic because they seem automatically wrong.

All scientists are prone to clinging to majorities despite counter evidence. Even scientists who have suffered mightily and unjustly for their minority opinions on some topics have themselves been irrationally hostile to minority opinions on others.

As Twain urged, it is essential to be aware of majorities and question yourself intensively when your views and practices align with them. One way to begin questioning is to study critiques of majority views and practices, such as those by scientists in a minority. Even critiques from the distant past may still be valid.

When we are in a minority, we at least doubt the majority, although we still likely suffer from insufficient doubt. Because researchers in the majority usually dismiss our work, we struggle to present and publish our ideas and evidence, and thus tend to receive little thoughtful and critical feedback. In a minority, we may also be professionally and socially marginal. Such isolation can lead to narrow thinking. The only remedies are to seek criticism from willing scientists (see Chapter 31) and intensify our efforts in doubting ourselves.

Extraordinary claims and extraordinary evidence

Sometimes scientists assert that "extraordinary claims require extraordinary evidence," especially when denigrating minority views or challenges to majority views. What counts as "extraordinary" and who decides? Often, scientists in the majority try to monopolize the right to classify what claims and evidence are indeed extraordinary. As Martin Lopez Corredoira noted, "… to make inconvenient changes in the worldview of the establishment, no evidence is extraordinary enough." Instead, simply evaluate all claims with evidence, along with other objective factors (see Chapter 58), by the same standards. Even without evidence, all testable claims—hypotheses—are worth considering and perhaps evaluating empirically.

Responding to doubters

Inquiry into the evidence of a doctrine is not to be made once for all, and then taken as finally settled. It is never lawful to stifle a doubt; for either it can be honestly answered by means of the inquiry already made, or else it proves that the inquiry was not complete.
—*William Kingdon Clifford, 1877*

... no idea should be suppressed ... and it applies to ideas that look like nonsense. We must not forget that some of the best ideas seemed like nonsense at first.
—Cecila Payne-Gaposchkin, 1979

Progress in science depends on researchers who challenge others' ideas and evidence. Doubters are scientific heroes, but researchers generally treat them as villains. Quite simply, those scientists who denounce and discourage doubters undermine the pursuit of truth gravely.

Doubts and questions are just ideas. Individual scientists, if they so choose, can think for themselves and evaluate any idea they encounter, including dissenting ones.

There are no necessary qualifications to be a constructive doubter. Scientists and non-scientists both can legitimately question research in any field. The principles of good scientific research, such as those highlighted in this book, do not require technical knowledge or scientific training to understand and therefore can be used by anyone in evaluating research. Beware of the scientist's defensive cry that a critic's doubt is invalid because the critic is not a specialist in the same field or not a scientist at all. In fact, critics outside of a field often provide the best criticism, as insiders may be blinded to their field's shortcomings due to self-interest and other factors.

Adopting a doubting and questioning attitude means not reflexively dismissing others just because their ideas seem outlandish. True doubters go further: they support other doubters by aiding their attempts to present their views and reinforcing their calls for logic and evidence. Such support does not imply agreement, rather it indicates commitment to free expression and rational debate (see Chapter 43). Interact with doubters socially and professionally just as you would anyone else. Isolating and dismissing dissenters is anti-scientific and discourteous. It takes courage to question the status quo, and the reaction from the scientific mob can be harsh. Even a few independent voices calling for reason and discourse may blunt or avert collective hostility to challengers.

Extending doubt to our personal lives

Several years have now elapsed since I first became aware that I had accepted, even from my youth, many false opinions for true, and that consequently what I afterwards based on such principles was highly doubtful; and from that time I was convinced of the necessity of undertaking once in my life to rid myself of all the opinions I had adopted, and of commencing anew the work of building from the foundation ...
—René Descartes, 1641

Once doubting and questioning become central to our daily scientific work, it is natural to apply them to other aspects of our lives, such as religious, political, and social beliefs. We came to accept most of these beliefs through non-scientific means, so they are inherently vulnerable to rational critique. Richard Feynman considered this extension of doubt to other realms one of the main values of science. Doubting and questioning, along with analogs to other essentials of scientific research, also are useful in routine activities, such as buying goods and services and handling legal, financial, medical, and other important matters.

Doubting and questioning *outside* of science is good practice for doubting and questioning *in* science. And it's easy practice in a way—all religious, political, and other ideologies are filled with inconsistencies, irrationality, and discordance with empirical evidence. The ideologies tend not to change in response to criticism or empirical evidence. In daily life, governments, businesses, the media, other organizations, and individuals make claims of fact based on standards of evidence weaker than in most scientific research, leaving them more prone to error, and bias especially. Non-scientific beliefs are potentially dangerous doubt suppressors that can bias your work on even tangentially related topics (see Chapter 48). Neutralizing the power of non-scientific ideologies can enhance your research in such cases.

A doubting mind may come at a price. Religious, political, and social beliefs and the institutions linked to them give meaning and purpose to many people. Expressing doubts and questions about these beliefs and institutions can make relationships with family and friends difficult. Most people want to *reduce* their doubt and uncertainty. Yet rational and empirical examination of any topic, scientific or otherwise, involves *increasing* doubt and uncertainty, at least in the beginning. If you express your doubts on non-scientific matters continually to others, they may interpret your doubting as negativity and personal criticism of themselves. If you encounter this reaction, keep doubting but limit the doubts you tell others.

Perhaps most importantly, doubting outside of science allows you to learn more than otherwise. If learning makes science fun, why not spread similar pleasure to the rest of your life with doubt?

62

Resisting Corruption

And the simple step of a simple courageous man is not to partake in falsehood, not to support false actions! Let THAT enter the world, let it even reign in the world—but not with my help.
—Aleksandr Solzhenitsyn, 1970

In my scientific career, ethical challenges have been far more daunting and prominent than the challenges of doing research itself. When I decided to become a scientist, I never anticipated that ethical struggles would be the main difficulties of doing scientific research.

I consider corruption in research to be any intentional deviation from achievable good scientific practice. Money and social influence are two primary proximate factors that can lead to corruption in scientific research.

Money

It is difficult to get a man to understand something when his salary depends on his not understanding it.
—Upton Sinclair, 1935

... the free university, historically the fountainhead of free ideas and scientific discovery, has experienced a revolution in the conduct of research. Partly because of the huge costs involved, a government contract becomes virtually a substitute for intellectual curiosity ... The prospect of domination of the nation's scholars by Federal employment, project allocations, and the power of money is ever present—and is gravely to be regarded.
—Dwight D. Eisenhower, 1961

But under nationalisation [of academic science funding], scientists need only relate to other scientists. Their money comes from

government grants which are judged and awarded by other scientists; government grants support the students; and scientists need only leave their laboratories to attend meetings (government assisted) with other scientists. No wonder researchers become narrow and intense and, often, moral cowards, terrified to offend. If you only inhabit one tiny world, and you are desperate for the recognition of that world, you are careful to be politically correct.
— *Terence Kealey, 1996*

... in modern conditions ... the inevitability of the victory of obvious scientific truth is a lot less evident ... it is not in anyone's self-interest to solve the problem ... who needs a solution to the 'problem' ...? — certainly not most of those engaged in the field! Certainly not the funding agencies, whose raise d'être is the continued confusion; certainly not the deans, chairmen and provosts whose insatiable hunger for overheads is thereby slaked; and of course, certainly not the PI's (sic) of all those contracts ... is it really true that more money = better science?
— *Philip Anderson, 2011*

Funders, employers, and clients sometimes pressure scientists to flout good scientific practice so as to further the sponsors' political, financial, or other interests. This pressure can be direct and explicit, such as requiring substandard or bad practices, or indirect and subtle, such as signaling to researchers that only certain outcomes are acceptable. Scientists risk losing their income if they do not comply.

Many researchers who depend on grants to support their research (and salaries) unwittingly can lose intellectual and ethical control of their work. The keen competition for funding leads to research proposals highly tuned to reviewers' and funders' wishes. Although the proposals may, in principle, be investigator-initiated, in reality they are frequently determined by what researchers think will get funded, as opposed to what they think might be most relevant, interesting, and sound. In these circumstances, there is no coercion; rather, scientists voluntarily corrupt themselves, to some extent, for the chance of funding.

Fortunately, there are several strategies for avoiding financial corruption. Scout potential funders, employers, and clients before working for them to assess their commitment to finding truth. Advocacy organizations and government agencies typically have missions to defend particular policies, world views, and services. These missions are like preordained conclusions that, from the organization's perspective, may not be challenged or debated, no matter the evidence.

Industrial firms may also be committed to specific policies, products, and services. However, to the extent they compete in a free market in which customers can easily judge the quality of their offerings, industrial firms have a vested interest in knowing the truth, at least for some aspects of their businesses, particularly for new products and services. The career and financial interests of employees, including executives, though, are not always aligned with their firms' best interests. Sometimes managers coerce researchers to produce particular results, or twist or ignore scientific evidence to further their own personal goals at the expense of the firm's integrity.

Contract research organizations and individuals who do research for hire often get caught in ethical quandaries. Their business depends on satisfying clients, even when they may care only about commercial concerns, publication, or political or policy agendas, instead of truth.

Academic research groups may have ideological conflicts of interest that hinder the pursuit of truth. Group leaders' commitments to theories, methods, interventions, and their own prior work may hinder open and free investigation and communication. At research universities, departments are most likely to hire and promote scientists who attract the most grant funding, follow current fashions in the field, and hew to research agendas set by senior, high status scientists. Universities may be bound to particular ideologies, policies, or sponsors so strongly that they disavow research done by their faculty and staff which conflicts with such agendas. Intellectually wayward faculty risk dismissal, despite nominal academic freedom.

Funding agencies guided by peer review are heavily biased toward conventional research that does not challenge the main hypotheses and dominant perspectives in a field. Many private foundations fund research that advances their ideological, political, or commercial interests only.

Moreover, some fields may be more likely to involve corrupting influences than others. Fields in which research funding overwhelmingly comes from a single funding source (usually government) are very susceptible to corruption. When government is the main research sponsor, it often indicates that research funding has political purposes.

Professions that are connected to scientific fields are also powerful political forces that shape the limits of acceptable research, regardless of specific funding source. In almost every case, the professions formed before any scientific basis existed for their practices. The scientific field effectively exists to serve the profession. Researchers thus might not question or investigate many of the assumptions underlying the profession. Practitioners' livelihoods might be endangered if scientists found the justifications for the profession or some of its major practices unwarranted in empirical research.

Unfortunately, most scientific workplaces and fields probably fit one or more of the preceding descriptions and thus are at high risk for corruption. So other strategies for avoiding financial corruption beyond scouting the landscape may be necessary.

A first response to corrupting pressure might be to make a case for good scientific practice. If your colleagues, bosses, sponsors, or clients are not persuaded, then you might see whether they would accept the good scientific practice as an addition to, rather than replacement of, what they want (if feasible). If this approach fails, then you could assess ethical matters on a project-by-project basis. Do not attach your name to any research you feel has been corrupted or does not meet your ethical standards, no matter how much or in what capacity you worked on it. Just consider that project a job for money only. Maybe other projects won't be compromised in this way.

If the ethical problems you face are more pervasive, seek other scientific employers, funders, or clients who might be less likely to apply corrupting pressures. If working in a particular field is no longer possible or tolerable, seek scientific employment, funding, or clients in other fields. Having prior research experience in other fields (see Chapter 49) increases the chance you will be able to switch successfully.

The most extreme response is to leave a professional career in scientific research. Most scientists have skills that can transfer to income-producing work related to research, such as teaching, consulting, editing, authoring books, programming and developing software, managing and analyzing (non-research) data, and many other non-research technical activities that involve scientific methods and knowledge. Or you can pursue a career entirely unrelated to science. Regardless, you can still do research in your free time without funding (see Chapter 16). If scientific research is what you love, it doesn't matter how you make a living as long as you get to contribute to science in some way. There is no shame in being mainly or solely employed outside of science. On the contrary, freedom in scientific research and peace of mind from keeping your integrity can make other sacrifices very worthwhile.

Some may criticize this view as being unrealistic—who would become a scientist without a good prospect of a professional career? To this criticism I respond that someone who is vulnerable to manipulation with money—either by character or circumstance—threatens the development of reliable scientific knowledge. Persons who are not deeply committed to seeking truth through scientific research cannot be counted on to produce trustworthy work. The quality of scientists matters far more than their quantity.

Social influence

... your researchers are free, but they are conditioned by the fashion of the day.
—Aleksandr Solzhenitsyn, 1978

In a healthy scientific field, researchers try to falsify hypotheses and theories. Challenges and challengers are welcomed. However, this ideal might not occur often or in most fields. Instead, researchers may guard reigning hypotheses and theories against attack. Sometimes scientists may even strive to protect dominant methods against competing techniques no matter their potential advantages. Such defensiveness inhibits the search for truth. When this mindset pervades a field, researchers are under great pressure to conform to mainstream viewpoints, which often leads them to neglect good scientific practice.

Although the most important reasons for doing scientific research do not concern others' reactions to our work (see Chapter 60), the reactions still affect us greatly. Most people want to be liked by their peers, or at least not be disliked by them. If you express unpopular or dissenting ideas in a scientific field afflicted with defensive attitudes, it is likely you will experience some degree of social conflict with or isolation from other researchers as a result. Although open debate about scientific matters is a scientific ideal, in practice, the differences aired in the scientific arena often spill over to social interactions outside of the debate.

If you get too far out of step with other researchers in your field, you could be branded as a heretic (or worse), harming your scientific reputation, mental health, and friendships with other researchers. When you dissent publicly, you become a liability to friends and colleagues who are in the mainstream of a field, as others may also judge them guilty by their association with you. Your dissent may also represent a potential financial threat to your mainstream friends, especially if your ideas and work are at odds with those promoted by major funders and your friends rely on those funders' support.

As shown by history and research, it takes remarkably little social pressure to get most individuals to conform to others' thoughts and behaviors, even against those individuals' better judgment and morals. Even the most stubborn and independent among us are vulnerable. So it is essential to have strategies to prevent social influence from corrupting your research.

One basic approach is to use social support to counter corrupting social influence. Seek others who have similar ideas and are working in the same vein as you, if there are any. Do this not to establish a competing but smaller pocket of groupthink, but to create a safe harbor from the social storm. You can have productive scientific discussions with such colleagues, which sometimes may evolve into collaborations. Furthermore, it is crucial to have good friends who are not involved in your controversy. Friends and family members who are not involved in science at all are a very important resource. Scientists working in other fields and other free-thinking scientists in your field who are outcasts on other topics also can be worthy partners for scientific discussion and other social interaction. As a result of being marginalized, we misfits may be more accepting of differences and eager to have friendly colleagues. All of these sources of social support enable you to be socially independent of those who might otherwise exert a corrupting influence on you.

Finally, you can take other steps to forestall social influence from corrupting your research. Do research without funding (see Chapter 16) and publish your reports on preprint servers (see Chapter 32) to remove so-called peer review as a limiting factor in your work. Replace coercive peer review with constructive feedback from other researchers that you seek directly (see Chapters 31 and 40) or receive spontaneously.

I close this section on ethical matters with the same hope that Richard Feynman once articulated:

... I have just one wish for you—the good luck to be somewhere where you are free to maintain the kind of integrity I have described, and where you do not feel forced by a need to maintain your position in the organization, or financial support, or so on, to lose your integrity. May you have that freedom.

Epilogue

The main obstacles to progress in most scientific fields are not technical — insufficient methods, computing power, or data. Nor are they in the complexity of nature. The barriers, rather, are in ourselves as scientists — our practices, mindsets, and economic and social organization. More funding and more scientists are not remedies; indeed, they could thwart progress. I believe that the essentials I highlight in this book are the keys to advancing science.

Unlike many people, we scientists produce work that may be kept and relied on for many generations. Our scientific contributions may be among our most lasting legacies. Remembering that fact may give us extra incentive to use all means available in our pursuit of truth.

Our time is forever growing shorter. So start a research project today, or continue an existing one with renewed vigor. And have fun on your journey of discovery!

Notes

Each note includes the page number for the corresponding text highlighted in bold.

Chapter 1. Introduction
1, Seneca quote: Seneca LA. On the philosopher's seclusion. In Seneca ad Lucilium epistulae morales, Gummere RM, translator. Vol. 1. London: William Heinemann, 1917. P. 37.

1, Leibniz quote: Leibniz GW. Discours touchant la méthode de la certitude et l'art d'inventer. Available at https://perma.cc/GD7G-CZQJ. Leibniz wrote this essay between 1688 and 1690. Translation by John Potterat.

1, Feynman quote: Feynman RP. The pleasure of finding things out. Cambridge, MA: Perseus Publishing, 1999. P. 211.

2, modern scientists quite often overlook these basic principles and practices: I wrote this book because I knew of no other like it. In the late stages of writing this book, however, I found two books that overlapped somewhat with mine—W. I. B. Beveridge's *The Art of Scientific Investigation* (W. W. Norton and Company, New York, 1957 [first published in 1950]) and Santiago Ramon y Cajal's *Advice for a Young Investigator* (translated by Neely and Larry W. Swanson, MIT Press, Cambridge, MA, 1999; based on the 4ᵗʰ edition of this book). The scope of both books is narrower than mine, and some of these authors' advice is no longer current (as surely some of mine will be, too, in time). Even though I composed my book without knowledge of these other books, for those topics Ramon y Cajal, Beveridge, and I all covered, our observations and advice are remarkably similar, with a few exceptions. Read both books for free online: *The Art of Scientific Investigation* at www.archive.org and *Advice for a Young Investigator* at http://www.webcitation.org/77C0MyLbk.

Chapter 2. No Substitute for Field Experience and Observation
9, Gascoyne-Cecil quote: From p. 303 of an article entitled "Lord Canning on missionary preaching" in the Saturday Review: Politics, Literature, Science and Art (September 10, 1859). The article was written anonymously, but Wikipedia (https://perma.cc/2AJD-T6SA) noted that he contributed articles anonymously to the Saturday Review in this period, and Wikiquote attributed an inaccurate version of this quote to him (https://perma.cc/NQ4T-2P68).

Chapter 3. Designing and Planning Research
12, Plato quote: Plato. Laws. 360 B.C. Translated by Jowett B. Available at https://perma.cc/4FB4-KTX2.
12, Cicero quote: Cicero MT. De officiis (On duties). Translated by Miller SW. Loeb edition. Cambridge, MA: Harvard University Press, 1913. P. 11.
12, Hofstadter quote: Hofstadter D. Gödel, Escher, Bach: an eternal golden braid. London: Penguin Books, 1979. P. 160.

Chapter 4. Pre-Registering Research
15, Garfield quote: Forty thousand quotations: prose and poetical. Compiled by Douglas CN. New York: Halcyon House, 1904. P. 1094.
16, several registries exist: For clinical trials, see the registries listed at the International Clinical Trials Registry Platform (http://apps.who.int/trialsearch/). The Open Science Framework (https://osf.io) and the Research Registry (http://www.researchregistry.com/) are two registries that serve scientists in all fields for all types of studies. There are also registries for particular fields, such as economics (https://www.socialscienceregistry.org/) and political science (http://egap.org/).

Chapter 5. Searching the Literature
18, Terentius (Terence) quote: Afer PT. Heautontimorumenos: the self-tormentor. Riley HT, editor. Act IV, scene 2, line 8.

Available at https://perma.cc/R75G-S8Z9. Written in 163 B.C., according to Wikipedia (https://perma.cc/5HCX-ZA9X).

18, John of Salisbury quote: John of Salisbury. The metalogicon of John of Salisbury: a twelfth century defense of the verbal and logical arts of the trivium. Translated by McGarry DD. Berkeley: University of California Press, 1955. P. 167 (book 3, chapter 4). Originally published in 1159.

18, Huxley quote: Huxley A. Proper studies: the proper study of mankind is man. London: Chatto and Windus. 1957. P. 205. First published in 1927.

19, PubMed: www.ncbi.nlm.nih.gov/pubmed

19, Google Scholar: scholar.google.com

20, Google Scholar captured the vast majority of the scientific literature: For example, see:

Khabsa M, Giles CL. The number of scholarly documents on the public web. PLoS One. 2014 May 9;9(5):e93949. Available at http://journals.plos.org/plosone/article?id=10.1371/journal.pone.0093949.

Gusenbauer M. Google Scholar to overshadow them all? Comparing the sizes of 12 academic search engines and bibliographic databases. Scientometrics. 2019;118:177–214.

Martín-Martín A, Thelwall M, Orduna-Malea E, López-Cózar ED. Google Scholar, Microsoft Academic, Scopus, Dimensions, Web of Science, and OpenCitations' COCI: a multidisciplinary comparison of coverage via citations. arXiv:2004.14329, 2020 April 29. Available at https://arxiv.org/abs/2004.14329.

21, Santayana quote: Santayana G. The life of reason or the phases of human progress. Volume 1, Reason in common sense. London: Archibald Constable and Co. Ltd., 1910. P. 284. First published in 1905.

Chapter 6. Collecting Data

22, Brackett quote: Brackett AC. The technique of rest. New York: Harper and Brothers, 1892. P. 77.

22, Schrödinger quote: Schrödinger E. What is life? The physical aspect of the living cell and mind and matter. Cambridge: Cambridge University Press, 1944. P. 176.

22, Beveridge quote: Beveridge WIB. The art of scientific investigation. New York: W. W. Norton and Company, 1957. P. 17. This book was first published in 1950.

Chapter 7. Managing Data
26, Asimov quote: Asimov I, Shulman JA (editors). Isaac Asimov's book of science and nature quotations. New York: Weidenfeld & Nicolson, 1988. P. 318. Asimov was a biochemist as well as a science fiction writer.

Chapter 8. Analyzing Data
28, Thomson quote: Thomson W. Electrical units of measurement. Popular lectures and addresses. Volume 1, Constitution of matter. London: Macmillan and Co., 1889. P. 73.
28, Meehl quote: Meehl PE. Theory-testing in psychology and physics: a methodological paradox. Philosophy of Science. 1967;34:103-115.

Chapter 9. Archiving and Sharing Data
37, motto of the Royal Society: See http://archive.ph/4HWRu.
37, Ramon y Cajal quote: Ramon y Cajal S. Advice for a young investigator. Translated by Swanson N and Swanson LW. Cambridge, MA: MIT Press, 1999. P. 86. Translation based on the 4th edition of Ramon y Cajal's book published in 1916.
38, research for which the raw data are unavailable ... bar such work from the literature: See Gramling C. "Four-legged snake" may be ancient lizard instead. Science. 2016; 354(6312): 536–537.
39, policies requiring authors and researchers to share data: In many fields during recent decades, few authors and funded researchers shared their data on request despite operating under such policies. For examples, see:
Wicherts JM, Borsboom D, Kats J, Molenaar D. The poor availability of psychological research data for reanalysis. American Psychologist. 2006;61(7):726.
Savage CJ, Vickers AJ. Empirical study of data sharing by authors publishing in PLoS journals. PLoS One. 2009;4(9):e7078.

Vines TH, Albert AY, Andrew RL, Débarre F, Bock DG, Franklin MT, Gilbert KJ, Moore JS, Renaut S, Rennison DJ. The availability of research data declines rapidly with article age. Current Biology. 2014;24(1):94–7.

Couture JL, Blake RE, McDonald G, Ward CL. A funder-imposed data publication requirement seldom inspired data sharing. PLoS One. 2018;13(7):e0199789.

40, general and field-specific scientific data archives: Some examples are GenBank, Addgene, Open Science Framework, Dataverse, Figshare, Zenodo, Dryad, and ICPSR. Many others exist.

Chapter 10. Using Secondary Data

43, Deutscher and colleagues quote: Deutscher I, Pestello FP, Pestello HFG. Sentiments and acts. New York: Aldine de Gruyter, 1993. P. 31.

44, many authors ignore or decline requests to share data: Over the years, I have made hundreds of requests for data from researchers in many different fields. Most researchers simply do not respond to such requests, even though they are still active in science (see Chapters 40 and 41). Only a small fraction ultimately ever share their data. My experience matches that of others who have requested data (see Notes for Chapter 9).

Chapter 11. Programming

46, Sayre quote: Sayre D. Panel discussion of "A new concept in programming" by Brown GW. In Greenberger M (Editor). Management and the Computer of the Future. Cambridge, MA: MIT Press and New York: John Wiley and Sons, Inc., 1962. P. 277.

46, Braben quote: Braben D. To be a scientist. Oxford: Oxford University Press, 1994. P. 60.

Chapter 12. Backing up Research

49, Nansen quote: Nansen F. Farthest north, volume II. New York: Harper & Brothers Publishers, 1897. P. 16.

Chapter 13. Keep it Simple, Scientist (KISS)

52, Newton quote: Quoted in Manuel FE. The religion of Isaac Newton. Oxford: Clarendon Press, 1974. P. 120.

52, Romney quote: Kim Romney said this to me in partial jest in the early 1990s.

53-54, Popper quote: Popper KR. The logic of scientific discovery. London: Routledge, 2005. P. 128. First published in 1935.

54, Whitehead quote: Whitehead AN. The concept of nature. The Tarner Lectures delivered in Trinity College November 1919. Cambridge: Cambridge University Press, 1964. P. 163.

Chapter 14. Replicating Research

57, Bernard quote: Bernard C. An introduction to the study of experimental medicine. Translated by Greene HC. New York: Henry Schuman, Inc., 1949. P. 91. This book was first published in 1865.

57, Pearson quote: Pearson K. The grammar of science. London: Adam and Charles Black, 1892. P. 16.

Chapter 15. Reviewing Systematically

60, Leibniz quote: Leibniz GW. Discours touchant la méthode de la certitude et l'art d'inventer. Available at https://perma.cc/5Q38-DFJK. Leibniz wrote this essay between 1688 and 1690. Translation by John Potterat.

62-63, Crichton quote: Crichton M. Aliens cause global warming. Caltech Michelin Lecture, January 17, 2003. Available at https://perma.cc/52RA-M4BJ. Crichton was not only a writer and filmmaker, but also a scientist and physician.

Chapter 16. Doing Research without Funding

64, Pearson quote: Pearson K. The grammar of science. London: Adam and Charles Black, 1892. P. 5.

64, Curie quote: Curie M. Pierre Curie. Translated by Kellogg C, Kellogg V. New York: Macmillan Company, 1923. P. 185.

64, Rutherford quote: As quoted by his student, Bertram Vivian Bowden. Bowden L. Expectations for science. New Scientist 1965;September 30:851.

64, Anderson quote: Anderson PW. More and different: notes from a thoughtful curmudgeon. Singapore: World Scientific Publishing Co. Pte. Ltd.; 2011. P. v.

66, Dharmadhikari's construction of highly productive scanning tunneling microscopes: Sekhsaria P. The making of an indigenous scanning tunneling microscope. Current Science 2013;104(9):1152–1158.

Chapter 17. Checking—Part I
68, Gawande quote: Gawande A. The checklist manifesto. New York, Metropolitan Books, 2009. P. 186.

68, anonymous quote: Gawande A. The checklist manifesto. New York, Metropolitan Books, 2009. P. 168.

68, The consistent use of checklists has dramatically improved safety and efficiency: Gawande A. The checklist manifesto. New York, Metropolitan Books, 2009.

Chapter 18. Simplifying the Writing Process
73, Pope quote: Pope A. Pope's essay on criticism. Edited by Collins JC. London: Macmillan and Co., Limited, 1896. P. 11. This poem was first published in 1711.

73, Becker quote: Becker HS. Writing for social scientists. Chicago: University of Chicago Press, 2007. P. 60. This book was first published in 1986.

75, Pile sorting is a method: See Weller SC, Romney AK. Systematic data collection. Newbury Park, California: Sage Publications, 1988.

75, Pile sorting also works well when writing on paper: Howard Becker described the same sorting process with text written on pieces of paper in the era before personal computers and word processors were common. See Becker HS. Writing for social scientists. Chicago: University of Chicago Press, 2007. This book was first published in 1986.

77, Many free (reference manager) **programs exist**: See https://en.wikipedia.org/wiki/Comparison_of_reference_manag ement_software.

Chapter 19. Writing a Methods Section
78, Kipling quote: Kipling R. The elephant's child. Just so stories for little children. New York: Doubleday, Page and Company, 1907. P. 83. First published in 1902.
80, digits from atmospheric noise: See www.random.org.
82, Internet Archive: www.archive.org
82, Open Science Framework: www.osf.io

Chapter 20. Writing a Results Section
83, Curie quote: Curie M. Pierre Curie. Translated by Kellogg C, Kellogg V. New York: Macmillan Company, 1923. Pp. 111, 226. Available at https://perma.cc/JYT6-HFSM and https://perma.cc/7SXC-JKLK.
84, another stable location: Open Science Framework (www.osf.io) and the Internet Archive (www.archive.org).

Chapter 21. Writing a Discussion Section
87, Popper quote: Popper K. The logic of scientific discovery. London: Routledge, 2002. P. 280. The book was first published in 1959.
87, Feynman quote: Feynman RP. The pleasure of finding things out. Cambridge, MA: Perseus Publishing, 1999. Pp. 209–210. From an address Feynman gave in 1974.

Chapter 22. Writing an Introduction
93, Hughes quote: Becker HS. Writing for social scientists. Chicago: University of Chicago Press, 2007. P. 50.
93, Jeffrey McDonnell's three questions to address: McDonnell, J. J. (2017). Paper writing gone Hollywood. Science. 2017;355(6320):102.

Chapter 23. Writing an Abstract

96, Ulam quote: See Rota G-C. Words spoken at the memorial service for S. M. Ulam (The Lodge, Los Alamos, New Mexico, May 17, 1984). The Mathematical Intelligencer. 1984;6(4):40–41.

Chapter 24. Writing Acknowledgments

98, Donne quote: Donne J. Devotions upon emergent occasions, meditation 17. 1623. Available at https://perma.cc/3AFK-ND93.
98, Colman quote: Colman G. The heir at law. London: Longman, Hurst, Rees, Orme, and Brown, Paternoster-Row, 1818. P. 8. This play was first published in 1797.

Chapter 25. Figuring

100, Cheysson quote: Ministere des Travaux Publics, Cheysson E. Notice. Direction des Cartes, Plans et Archives, et de la Statistique Graphique. Album de Statistique Graphique. Imprimerie Nationale, Paris, July, 1880. P. vi. Available at https://perma.cc/PJ5X-2R7D. Translation mostly from Sigal M, Friendly M. Some prehistory of CARME: visual language and visual thinking. In Visualization and verbalization of data, (Blasius J, Greenacre M, editors). Boca Raton: CRC Press, 2014. P. 44. I thank Michael Friendly for his help in finding the original source.
102, Graphical display strategies vary dramatically in their effectiveness: Cleveland WS, McGill R. Graphical perception: theory, experimentation, and application to the development of graphical methods. Journal of the American Statistical Association. 1984;79(387):531–54.

Chapter 26. Citing Sources

105, Seneca quote: Seneca LA. On tranquility of mind. Basore JW, translator, Loeb Classical Library. London: William Heinemann, 1928–1935. Available at https://perma.cc/3JT5-UZEA.
105, Bayle quote: Reportedly in Bayle's second volume of Works, as quoted in Hoyt JK. Hoyt's new cyclopedia of practical

quotations (revised by Roberts KL). New York: Funk and
Wagnalls Company, 1922, p. 653.

106, the author(s) popularly perceived as the first: Stigler's Law
of Eponymy captures a similar phenomenon: "no scientific
discovery is named after its original discoverer." Stigler S.
Stigler's law of eponymy. Transactions of the New York
Academy of Sciences. 1980;39(1 Series II):147–158.

**107, authors not actually reading the sources they cite, a
practice which unfortunately may be widespread**: Based on the
pattern and frequency of misprinted citations, Simkin and
Roychowdhury estimated that the large majority of highly cited
articles in physics journals had not been read by the citing
authors (Simkin MV, Roychowdhury VP. Stochastic modeling of
citation slips. Scientometrics. 2005;62(3):367–384.).

Chapter 27. Using Accurate Language

109, Twain quote: From Twain's letter to George Bainton, written
on October 15, 1888, and solicited for and printed in George
Bainton's *The Art of Authorship: Literary Reminiscences, Methods of
Work, and Advice to Young Beginners* (1890, D. Appleton and
Company, New York, pp. 87–88).

Chapter 28. Writing across Generations

112, Yang quote: Needham J. Science and civilization in China:
volume 3, mathematics and the sciences of the heavens and the
Earth. Cambridge: Cambridge University Press, 1959. P. 104.

112, Kepler quote: Kepler J. Harmonices mundi (The harmony
of the world). Translated by Aiton EJ, Duncan AM, Field JV.
Philadelphia: American Philosophical Society, 1997. P. 391. First
published in 1619.

112, Channing quote: Channing WE. Self culture. Speech
delivered in Boston, USA, September, 1838. Available at
https://perma.cc/EC9N-ZW3Q.

112, writing for scientists of your grandparents' generation: As
much as possible, use words only if they appear in relatively old
unabridged dictionaries. The Internet Archive

(www.archive.org) has several unabridged dictionaries from the late 1800s.

Chapter 29. Less is More

114, Thales quote: Harbottle TB. Dictionary of quotations (classical). London: Swan Sonnenshein and Co., Limited, 1906. P. 455.

114, Shakespeare quote: Shakespeare W. Hamlet. London: Cassell and Company, Limited, 1899. P. 64 (Act 2, scene 2).

114, Ruskin quote: Ruskin J. A joy for ever. London: George Allen, 1906. P. 188. First published in 1857.

117, Howard Becker's method: Becker HS. Writing for social scientists. Chicago, University of Chicago Press, 2007.

119, Pascal quote: Pascal B. Letter XVI, to the reverend fathers, the Jesuits. The provincial letters. Translated by McRie T. Edinburgh: John Johnstone, 1847. P. 282. Pascal wrote this letter on December 4, 1656.

120, Einstein quote: From a translation of Einstein's Autobiographische Skizze, in Stachel J. Einstein from 'B' to 'Z'. Boston: Birkhaeuser, 2001. P. 5.

For other tips about how to make your writing simpler and more direct, see: Pinker S. Why academics stink at writing. The Chronicle of Higher Education. September 26, 2004. Available at https://perma.cc/HS95-DSHP.

Chapter 30. Checking—Part II

122, Routh quote: Spoken by Martin Joseph Routh to John William Burgon on November 29, 1847. See Burgon JD. Lives of twelve good men. Vol. 1, third edition. London: John Murray, 1889. P. 73.

122, Safire quote: Safire W. On language. New York: Times Books; 1980. P. 100. Safire omitted the word "left" intentionally for humor.

122, In some fields, there are common checklists for reporting: See, for instance, www.equator-network.org and https://www.nc3rs.org.uk/arrive-guidelines.

Chapter 31. Soliciting Comments
124, Fuller quote: Fuller T. The holy state and the profane state. London: Thomas Tegg, 1841. This book was first published in 1642.
124, Bernard quote: Bernard C. The cahier rouge of Claude Bernard, translated by Hoff HH, Guillemin L, Guillemin R. Cambridge, MA: Schenkman Publishing Company, Inc., 1967. P. 221. Bernard made his notes in this book between 1850 and 1860.
124, Popper quote: Popper K. On freedom. In Popper K. All life is problem solving, translated by Camiller P. London: Routledge, 1999. P. 84. Popper first presented this essay in 1958.
125, discuss the problem with a colleague: For a more in-depth review of this topic, see Beveridge W. I. B. The art of scientific investigation. New York: W. W. Norton and Company, 1957.

Chapter 32. Publish, Then Perish
128, Deuteronomy Rabbah quote: Rapaport S. Tales and maxims from the Midrash. London: George Routledge and Sons Limited, 1907. P. 151. Scholars estimate that the Deuteronomy Rabbah was originally written sometime between AD 650 and 900 (https://perma.cc/4BP4-N7JH).
128, Leibniz quote: Leibniz GW. Discours touchant la méthode de la certitude et l'art d'inventer. Available at https://perma.cc/GD7G-CZQJ. Leibniz wrote this essay between 1688 and 1690. Translation by John Potterat.
128, Faraday quote: As quoted in Anonymous. Faraday. The Chemical News and Journal of Physical Science. 1867;16(404):111.
128, Nansen quote: Nansen F. Farthest north. New York: Harper & Brothers Publishers, 1897. P. 7.
128, Leonardo quote: da Vinci L. The notebooks of Leonardo da Vinci. Translated by Richter JP. 1888. XXI Letters. Personal records. Dated notes. Text number 1365. Available at https://perma.cc/L4P6-89ZN.
129, Feynman quote: Feynman RP. The pleasure of finding things out. Cambridge, MA: Perseus Publishing, 1999. P. 212. From an address Feynman gave in 1974.

131, Lord Acton quote: Lord Acton. A lecture on the study of history. London: Macmillan and Co., Limited, 1911. P. 62. From a lecture Acton gave in 1895.

131, Oppenheimer quote: Oppenheimer JR. Encouragement of science. Science News Letter. 1950;57(11):170–172.

132, Preprint servers: The number of preprint servers is continually growing, as are the services they offer. Two of the largest and oldest preprint servers are arXiv.org and ssrn.com. Open Science Framework Preprints (osf.io/preprints/) is a preprint server and clearinghouse for several other preprint servers. Many preprint servers unfortunately inhibit open scientific communication by blocking submissions from certain classes of authors and/or screening submissions. Preprint servers that screen submissions may reject papers on arbitrary non-scientific criteria, sometimes even beyond their stated policies. Screening is, in effect, like the censorship practiced by traditional journals. Some preprint servers even blacklist particular authors. As of May 2020, Open Science Framework Preprints, Zenodo (zenodo.org), and Figshare (figshare.com) did not screen submissions or blacklist authors. Although not a preprint server, the Internet Archive (archive.org) also has no such censorship and is another alternative for publishing.

Chapter 33. Giving a Scientific Presentation

135, Freeman quote: Lin Freeman said this to me in the early 1990s.

135, A presentation is an advertisement: I learned this from Barry Wellman. See: Wellman B. Bum raps: daydreams of a weary conferencer. Footnotes. 1993;21(5):14. Available at https://perma.cc/CEV4-DUTC.

142, **evidence is very inconsistent about the merit of including text-oriented visual aids**: Adesope OO, Nesbit JC. Verbal redundancy in multimedia learning environments: a meta-analysis. Journal of Educational Psychology. 2012;104(12):250–263.

Chapter 34. Dealing with the Press
145, Jefferson quote: Jefferson T. Letter to John Norvell, June 11, 1807. Available at https://archive.ph/bUSZ.
145, Feynman quote: Feynman RP. The pleasure of finding things out. Cambridge, MA: Perseus Publishing, 1999. P. 212. From a lecture Feynman gave in 1974.
145, Kubrick quote: Kubrick on Barry Lyndon: an interview with Michel Ciment. Available at http://archive.ph/9MLbb. Vladimir Nabokov actually handled interviews differently: "The interviewer's questions have to be sent to me in writing, answered by me in writing, and reproduced verbatim" (Nabokov V. Strong opinions. New York: McGraw-Hill Book Company, 1973. P. xi.). Kubrick's humorous account and Nabokov's approach both illustrate attempts to prevent misrepresentation.

Chapter 35. Working with Other Researchers
151, English proverb:
https://www.phrases.org.uk/meanings/proverbs.html.
151, Beveridge quote: Beveridge WIB. The art of scientific investigation. New York: W. W. Norton and Company, 1957. P. 123. This book was first published in 1950.
152, Large teams …. Decreased independence of thought is one major drawback: See Wu L, Wang D, Evans JA. Large teams develop and small teams disrupt science and technology. Nature. 2019;566:378–382, and sources cited therein.
155, Idea bandits: See Beveridge WIB. The art of scientific investigation. New York: W. W. Norton and Company, 1957. P. 145. This book was first published in 1950.

Chapter 36. Recognizing Scientific Skills, Knowledge, and Character
161, Eriugena quote: Eriugena JS. Periphyseon I (de divisione naturae). Edited and translated by Sheldon-Williams IP. Dublin: Dublin Institute for Advanced Studies, 1968. Book 1, chapter 69. As cited in Dronke P (editor). A history of twelfth-century

Western philosophy. Cambridge: Cambridge University Press, 1988. P. 8.

161, Shakespeare quote: Shakespeare W. Shake-speare's sonnets: never before imprinted. London: Thomas Thorpe, 1609. Sonnet 69.

161, Levi-Montalcini quote: D'Emilio F. Nobel-winning biologist Rita Levi-Montalcini dies at 103. Associated Press, December 30, 2012. Available at https://perma.cc/HTH7-E94U.

163, Ernest Everett Just: See Manning KR. Black Apollo of science: the life of Ernest Everett Just. New York: Oxford University Press, 1983.

Chapter 37. Working with Research Partners
165, **Nightingale** quote: Nightingale E. Lead the field. Shippensburg, PA: Sound Wisdom, 2018. P. 18. This book was first published in 1966.

Chapter 38. Managing vs. Doing Research
167, Faraday quote: Faraday M. Letter to Christopher Hansteen. December 16, 1857. In James FAJL (editor). The correspondence of Michael Faraday. Volume 5, 1855–1860. Herts, UK: The Institution of Engineering and Technology, 2008. P. 314.

167, Topalidou quote: Topalidou I. The freedom of choice. Science. 2018;359(6382):1434.

Chapter 39. Handling Authorship
170, Crick quote: Crick F. What mad pursuit. New York: Basic Books, 1988. P. 135.

Chapter 40. Contacting Other Scientists
175, Confucius quote: Confucius. The analects. Translated by Legge J. Chapter 7 (Shu Er), paragraph 22. Available at http://ctext.org/analects/shu-er. Many scholars believe "The analects" were written between 480 and 350 B.C.

175, Braley quote: Braley B. Do it now. 1915. Available at https://perma.cc/9ZEE-LMLW. Wikipedia indicates this poem was first published in 1915 or earlier (https://perma.cc/7K6P-XQLM).

177, relevant social media: As of 2020, some of the most relevant social media services are ResearchGate, Facebook, and LinkedIn.

Chapter 41. Responding to Other Scientists

179, Stanhope quote: Stanhope P. Letters to his son by the Earl of Chesterfield on the fine art of becoming a man of the world and a gentleman. New York: M. Walter Dunne, 1901. P. 85. The quote is from a letter Stanhope wrote on July 1, 1748.

179, Curie quote: Curie M. Pierre Curie. Translated by Kellogg C, Kellogg V. New York: Macmillan Company, 1923. P. 111. Available at https://perma.cc/FR2Q-252X.

Chapter 42. Giving Comments

181, Mill quote: Mill JS. On liberty. London: Watts and Co., 1929. P. 20. First published in 1859.

181, Pauling quote: Pauling L. No more war. New York: Dodd, Mead & Company, 1958. P. 209.

181, Potterat quote: John Potterat said this to me on several occasions beginning in the early 2000s.

182, matter and manner: I follow John Potterat's elegant phrasing of these two components.

Chapter 43. Engaging in Scientific Debates

185, Herbert quote: Herbert G. The church-porch. In The temple. London: Methuen & Co., 1899. P. 16. This book was first published in 1633.

185, first Popper quote: Popper K. The myth of a framework: in defence of science and rationality. Notturno MA (Editor). London: Routledge, 1994. P. 94. The quoted work was first published in 1963.

185, second Popper quote: Popper K. The myth of a framework: in defence of science and rationality. Notturno MA (Editor). London: Routledge, 1994. Pp. 37, 44. The quoted work was first published in 1976.

186, Sophocles quote: Harbottle TB. Dictionary of quotations (classical). London: Swan Sonnenshein and Co., Limited, 1906. P. 453.

191, post-publication peer review site: As of 2020, a popular site is pubpeer.com.

Chapter 44. Chairing a Conference Session
192, Bacon quote: Bacon F. Essays, civil and moral. Of discourse, No. 32. London: A. Swalle and T. Childe, 1696. P. 90. First published in 1597. Available at https://perma.cc/A4WH-V5Q9.

Chapter 45. Chasing Wild Geese
199, Bernard quote: Bernard C. The cahier rouge of Claude Bernard. Translated by Hoff HH, Guillemin L, Guillemin R. Cambridge, MA: Schenkman Publishing Company, Inc., 1967. P. 84. Bernard made his notes in this book between 1850 and 1860.

199, Pauling quote: As quoted by Francis Crick in his 1995 presentation, "The impact of Linus Pauling on molecular biology," available at https://perma.cc/6RST-XGBG.

199, Bronowski quote: Bronowski J. The reach of imagination. American Scholar 1967;36(2):199.

Read W. I. B. Beveridge's *The Art of Scientific Investigation* (New York: W. W. Norton and Company, 1957; available at www.archive.org) for many interesting accounts of how scientists stimulate and experience their imaginations and intuitions.

Chapter 46. Making Notes
202, Darwin quote: Darwin C. The life and letters of Charles Darwin. Darwin F, editor. New York: D. Appleton and Company, 1897. P. 71.

202, Aqua Notes motto: https://perma.cc/2WHW-R2J9.

202, Leonardo da Vinci making notes: da Vinci L. The notebooks of Leonardo da Vinci. Translated by Richter JP. 1888. Chapter 9 ("The practice of painting"), section 571. Available at https://perma.cc/74B2-4NEB.

Chapter 47. Reading the Literature

204, Ecclesiasticus quote: Available at https://perma.cc/47WN-AVYC.

204, John of Salisbury quote: From Pike JB. Frivolities of courtiers and footprints of philosophers: being a translation of the first, second, and third books and selections from the seventh and eighth books of the Policraticus of John of Salisbury. New York: Octagon Books, 1972. P. 224. Available at https://perma.cc/BU7B-QNYQ. According to Wikipedia, John of Salisbury wrote Policraticus in 1159 (http://archive.vn/dcRon).

204, Pearson quote: Pearson K. The grammar of science. London: Adam and Charles Black, 1892. P. 16.

204, Einstein quote: Einstein A. Letter to Robert A. Thornton, December 7, 1944. Einstein Archives, Hebrew University of Jerusalem, archival call number 56-283. Available at http://alberteinstein.info/vufind1/Record/EAR000041430. © The Hebrew University of Jerusalem. Quoted with permission.

207, high status journals publish less reliable and less rigorous research: See Brembs B. Prestigious science journals struggle to reach even average reliability. Frontiers in Human Neuroscience. 2018;12:37.

Chapter 48. Choosing Productive Scientific Problems and Questions

210, Herschel quote: Herschel JFW. A preliminary discourse on the study of natural philosophy. Philadelphia: Carey and Lea, 1831. P. 11.

210, first Feynman quote: Feyman RF. The pleasure of finding things out. Cambridge, MA: Perseus Publishing, 1999. P. 203. From an interview article published by Omni Magazine in 1992.

211, Popper quote: Popper K. Conjectures and refutations. London: Routledge, 2002. P. 173. From a lecture Popper first gave in 1948.

211, second Feynman quote: Feynman RP. The pleasure of finding things out. Cambridge, Massachusetts: Perseus Publishing; 1999. P. 104. From a lecture Feynman gave in 1964.

213, Whitehead quote: Whitehead AN. Objects and subjects. Proceedings and Addresses of the American Philosophical Association. 1931;5:132.

213, Braben quote: Braben D. To be a scientist. Oxford: Oxford University Press, 1994. P. 153.

213, Gould quote: Gould SJ. Full house: the spread of excellence from Plato to Darwin. New York: Harmony Books, 2011. Pp. 212–213. This book was first published in 1996.

Chapter 49. Crossing Disciplinary Boundaries

218, Bernard quote: Bernard C. The cahier rouge of Claude Bernard. Translated by Hoff HH, Guillemin L, Guillemin R. Cambridge, MA: Schenkman Publishing Company, Inc., 1967. Pp. 138–139. Bernard made his notes in this book between 1850 and 1860.

218, Thompson quote: Thompson SP. Calculus made easy, second edition. New York: Macmillan Company, 1914, p. xi. The first edition of this book was published in 1910.

218, Popper quote: Popper KR. The myth of a framework: in defence of science and rationality. Routledge: London, 1994. P. x.

219, Bacon quote: Bacon R. The "Opus majus" of Roger Bacon. Bridges JH (editor). Oxford: Clarendon Press, 1897. P. ix.

219-220, Tarzan complex: See Anderson PW. More and different: notes from a thoughtful curmudgeon. Singapore: World Scientific Publishing Co Pte Ltd.; 2011. P. 212.

Chapter 50. Abolishing the Law of the Method

221, Kaplan quote: Kaplan A. The conduct of inquiry. Jaipur, India: Sachin Publications, 1980. P. 28. This book was first published in 1964.

Chapter 51. Hypothesis Evaluation, Description, and Exploration

223, first Feynman quote: Feynman RP. The meaning of it all: thoughts of a citizen-scientist. Reading, Massachusetts: Perseus Books, 1998. P. 21. From a lecture Feynman gave in 1963.

223, second Feynman quote: Feynman RP. The meaning of it all: thoughts of a citizen scientist. Reading, Massachusetts: Perseus Books, 1998. P. 81. From a lecture Feynman gave in 1963.

223, activities that together cover most empirical scientific research: Development of theory, development of method, commentary, and speculation are scientific activities not included in these categories.

226, mountain that looked like a face: See Webb S. If the universe is teeming with aliens … where is everybody? Fifty solutions to the Fermi paradox and the problem of extraterrestrial life. New York: Copernicus Books, 2002. Pp. 40–41.

230, descriptive and exploratory research do involve hypotheses of a sort: See Popper K. Conjectures and refutations. London: Routledge, 2002.

Chapter 52. No Substitute for Experimental Control
231, Jabir quote: Quoted without citation in Holmyard EJ. Makers of chemistry. Oxford: Clarendon Press, 1931. P. 79.

231, Bernard quote: Bernard C. The cahier rouge of Claude Bernard. Translated by Hoff HH, Guillemin L, Guillemin R. Cambridge, Massachusetts: Schenkman Publishing, 1967. P. 37. Bernard made his notes in this book between 1850 and 1860.

237, True experiments are also possible … even when there is only a single case: N-of-1 trial designs (also known as reversal designs) allow researchers and others to extend experimental control to many settings. See Janosky JE, Leininger SL, Hoerger MP, Libkuman TM. Single subject designs in biomedicine. Dordrecht: Springer Science & Business Media, 2009, and Kazdin AE. Single-case research designs: methods for clinical and applied settings. New York: Oxford University Press, 2011.

Chapter 53. You are Not Your Hypotheses or Results
238, Darwin quote: From a letter to T. H. Huxley on July 9, 1857. Darwin F (editor). More letters of Charles Darwin, volume 1. London: John Murray, 1903. P.98.

238, Bernard quote: Bernard C. An introduction to the study of experimental medicine. Translated by Greene HC. New York: Henry Schuman, Inc., 1949. Pp. 23, 38. This book was first published in 1865.

238, Beveridge quote: Beveridge WIB. The art of scientific investigation. New York: W. W. Norton and Company, 1957. Pp. 46–47. This book was first published in 1950.

238, Feynman quote: Feynman RP. The relation of science and religion. Engineering and Science (California Institute of Technology). 1956; June: 21.

238, Popper quote: Popper KR. Conjectures and refutations. London: Routledge, 2005. P. 329. This book was first published in 1962.

239, Chamberlain quote: Chamberlin TC. Studies for students: the method of multiple working hypotheses. Journal of Geology. 1897;5(8):847. Chamberlain first presented his essay in 1892.

Chapter 54. Simulations are ... Simulations
242, Twain quote: Twain M. Autobiography of Mark Twain, volume 3. Griffin B, Smith HE, Fischer V, Frank MB, Gagel A, Goetz SK, Myrick LD, Ohge CM (editors). Oakland, California: University of California Press, 2015. P. 99.

242, Einstein quote: Einstein A. On the generalized theory of gravitation. Scientific American. 1950;182(4):17.

242, Anderson quote: Anderson PW. More and different: notes from a thoughtful curmudgeon. Singapore: World Scientific Publishing Co Pte Ltd; 2011. P. 102.

242, Computer models offer some advantages over other kinds of models: Simulations can also be very useful for understanding the behavior of statistical tests and measures. However, statistical methods are abstract tools with known and finite elements. In contrast, for empirical phenomena, all of the factors involved and their relationships may not be known.

244, validation: I use this term only because researchers who evaluate models often use it. It is an inaccurate term; comparison is a more accurate term (see Chapter 27).

Chapter 55. Analysis for Understanding
247, Toulmin and Goodfield quote: Toulmin S, Goodfield J. The fabric of the heavens: the development of astronomy and dynamics. Chicago: University of Chicago Press; 1999. P. 41. This book was first published in 1961.
247, this knowledge did not include, for the most part, understanding: Human tracking of animals, as in hunting, is a significant exception. As performed by hunter-gatherers, at least, this skill represents a kind of hypothetico-deductive scientific inquiry, in which speculation about and evaluation of causes are critical. See Liebenberg L. The origin of science. Cape Town: CyberTracker, 2013; www.cybertracker.org.

Chapter 56. Reasoning
249, Galileo quote: Galilei G. Third letter on sunspots, from Galileo Galilei to Mark Welser, December 1, 1612. Translated by Drake S, in Discoveries and opinions of Galileo. Garden City, NY: Doubleday and Company, Inc., 1957. P. 134.
249, von Helmholz quote: von Helmholz H. Über das Verhältniss der Naturwissenschaften zur Gesammtheit der Wissenschaft. Rede zum Geburtsfest des höchstseligen Grossherzogs Karl Friedrich von Baden und zur akademischen Preisvertheilung am 22 November 1862. Heidelberg: George Mohr, 1862. P. 23. Translation from http://archive.ph/L8tjJ.
249, Chesterton quote: Chesterton GK. Novels on the Great War. In Collected works, volume 35. San Francisco: Ignatius Press, 1991. P. 293. First published in the The Illustrated London News, April 19, 1930.
249, Anderson quote: Anderson PW. More and different: notes from a thoughtful curmudgeon. Singapoere: World Scientific Publishing Co Pte Ltd.; 2011. P. 391. From a lecture Anderson first gave in 1983.
249, Children use scientific practices and principles naturally: See Gopnik A. Scientific thinking in young children: theoretical advances, empirical research, and policy implications. Science. 2012;337(6102):1623–7.

249, people have relied on scientific reasoning to survive, such as when tracking animals in hunting: See Liebenberg L. The origin of science. Cape Town: CyberTracker, 2013; http://www.cybertracker.org.

250, extensive list of fallacies: https://en.wikipedia.org/wiki/List_of_fallacies.

Chapter 57. Considering the Unseen

255, Bastiat quote: From Bastiat F. That which is seen, and that which is not seen. 1850. Many English copies of this work are available for free online without credit to either a translator or publisher. My source is available at https://perma.cc/A4VM-KMG9.

255, Feynman quote: Feynman RP. The meaning of it all: thoughts of a citizen-scientist. Reading, Massachusetts: Perseus Books, 1998. P. 23. From a lecture Feynman gave in 1963.

Chapter 58. Judging Hypotheses

257, Ramon y Cajal quote: Ramon y Cajal S. Advice for a young investigator. Translated by Swanson N and Swanson LW. Cambridge, MA: MIT Press, 1999. P. 117. Translation based on the 4th edition of Ramon y Cajal's book published in 1916.

257, Sowell quote: Sowell T. A conflict of visions: ideological origins of political struggles. Revised edition. New York: Basic Books, 2007. P. 6. This book was first published in 1987.

257, Crick/Watson quote: Crick F. What mad pursuit: a personal view of scientific discovery. New York: Basic Books, 1988. P. 60.

258, Popper quote: Popper KR. The logic of scientific discovery. London: Routledge, 2005. P. 273. This statement was added in the 1959 edition of the book.

259, Beveridge quote: Beveridge WIB. The art of scientific investigation. New York: W. W. Norton and Company, 1957. P. 47. This book was first published in 1950.

260, Ptolemy quote: Ptolemaeus C. Ptolemy's Almagest. Book III, section 1, On the length of the year. Toomer GJ, translator. London: Gerald Duckworth & Co. Ltd., 1984. P. 136.

261, Heraclitus quote: Heraclitus. Fragments. Translated by John Burnet. 1912. Available at https://perma.cc/M43E-L3ZA.
261, Darwin quote: Darwin C. The life and letters of Charles Darwin. Volume 1. Darwin F, editor. London: John Murray, 1887. Pp. 103–104. Darwin wrote this statement in 1881.
261, Hoyle quote: As quoted in Edge DO, Mulkay MJ. Astronomy transformed. New York: Wiley, 1976. P. 432.

Chapter 59. Curbing Scientific Chauvinism
264, Bowden quote: Bowden L. Expectations for science. New Scientist 1965;September 30:853.
264, Popper quote: Popper KR. The moral responsibility of the scientist. Bulletin of Peace Proposals. 1971;2(3):279.
264, In *The Economic Laws of Scientific Research* (1996, Macmillan Press Ltd., Basingstoke), Terence Kealey reviewed the relationships between technology and science and the role of funding in scientific research. Nicholas Rescher also highlighted how progress in science depends on technological advances (Rescher N. Scientific progress: a philosophical essay on the economics of research in natural science. Oxford: Basil Blackwell, 1978). Many others have since observed the same.
265, Scientists (and their institutions and funders) frequently oversell the potential practical applications of their work ... scientists' ideas for potential applications also tend to be uneconomic:
For a recent overview, see Funk J. What's behind technological hype? Issues in Science and Technology. 2019;36(1):36–42. Available at http://archive.ph/Y7uWP.

Chapter 60. Getting Immediate Rewards from Doing Research
268, al-Biruni quote: From al-Biruni's Treatise on Astrological Lots, as translated by Haddad FI, Pingree D, Kennedy ES. Al-Biruni's Treatise on Astrological Lots. In Kennedy ES (editor), Astronomy and astrology in the medieval Islamic world. Aldershot: Ashgate, 1998. I thank Frederick Starr for pointing me to this reference, which he quoted in his book, Starr SF. Lost enlightenment: Central Asia's golden age from the Arab

conquest to Tamerlane. Princeton, NJ: Princeton University Press, 2013. Pp. 279–280. The quote I use includes clarifications added by Haddad and colleagues and Starr.

268, van Leeuwenhoek quote: van Leeuwenhoek A. Letter to Professors Cink, Narez, Rega, and the other Gentlemen of the College of *The Wild Boar*. June 12, 1716. As quoted in Dobell C. Antony van Leeuwenhoek and his "Little Animals." New York: Harcourt, Brace and Company, 1932. Pp. 82–83.

268, Smith quote: Smith T. Letter from Dr. Theobald Smith. Journal of Bacteriology. 1934;27(1):20.

268, Franklin quote: As told to her colleague, Aaron Klug (https://perma.cc/4TTT-38NP).

270, political and bureaucratic forces work against actual application: Research funding decisions by government agencies, especially in the United States, are mostly determined by the opinions of other researchers through so-called peer review. As a result, many research projects may not have specific or even general support from either elected or unelected government officials. In fact, some officials may even oppose the funded research and the motivation for it. Other government officials may not necessarily oppose the research, but simply disregard it entirely. Still other government officials may support the research, but for diverse reasons. The funded research sometimes may be used as a substitute for political or bureaucratic action, to placate political friends and foes alike. In other cases, government officials, regardless of political affiliation, may use some funded research as a bludgeon against their adversaries, but deliberately ignore other funded research. These dynamics make it unlikely that government-funded research will be applied, given the political and bureaucratic divisions within government (the latter of which may have little to do with political ideology).

Government agencies from local to national levels may even invite scientists to discuss their research and how to apply it. Based on my experience and that of my colleagues and others, the typical outcome is that the research never gets applied by the agencies, due to a variety of possible reasons.

Undoubtedly, there are examples of government-funded research being applied. But these represent a very small fraction of all government-funded research. Some government-funded research that does get applied reflects prevailing political inclinations that preceded the research and the research provides cover to implement the political decision. Similarly, other government-funded research, independent of scientific merit, gets applied by government only after intense lobbying by the researchers involved or their advocates. And many, perhaps most, instances of application involve non-governmental organizations—usually for-profit companies—taking the initiative. Nonetheless, the application of government-funded research is still highly improbable.

270, Bhabha quote: Dhirubhai Ambani Institute of Information and Communication Technology. Quotations by 60 greatest Indians. https://perma.cc/NHA4-9RNM.

271, Nansen quote: Nansen F. Farthest north, volume II. New York: Harper and Brothers Publishers, 1897. P. 9.

Chapter 61. Doubting and Questioning

272, Abelard quote: From the prologue to Abelard's *Sic et Non* as translated by Frank Pierrepont Graves in *A History of Education During the Middle Ages and the Transition to Modern Times* (New York: The Macmillan Company, 1910. P. 53).

272, Bernard quote: Bernard C. An introduction to the study of experimental medicine. Translated by Greene HC. New York: Henry Schuman, Inc., 1949. Pp. 35–36. This book was first published in 1865.

272, first Payne-Gaposchkin quote: Payne-Gaposchkin C. Cecilia Payne-Gaposchkin: an autobiography and other recollections, second edition. Edited by Haramundanis K. Cambridge: Cambridge University Press, 1996. Pp. 232–233. Payne-Gaposchkin first published this book in 1979.

272, Sowell quote: Sowell T. Words that replace thought. Creators.com, May 17, 2013. Available at http://archive.vn/BULtj.

274, Highly intelligent persons may be more likely to engage in deception and self-deception: See Trivers R. The folly of fools. New York: Basic Books, 2011.

274, Leonardo quote: da Vinci L. The notebooks of Leonardo da Vinci. Edited by Richter IR. Oxford: Oxford University Press, 1952. P. 283.

274, first Ramon y Cajal quote: Ramon y Cajal S. Advice for a young investigator. Translated by Swanson N and Swanson LW. Cambridge, MA: MIT Press, 1999. Pp. 121–122. Translation based on the 4th edition of Ramon y Cajal's book published in 1916.

274, first Feynman quote: Feynman RP. The pleasure of finding things out. Edited by Robbins J. Cambridge, MA: Perseus Publishing, 1999. P. 212. From a lecture Feynman gave in 1974.

275, why Charles Darwin took 20 years to publish *On the Origin of Species*: Darwin C. Charles Darwin: his life told in an autobiographical chapter. Edited by Darwin F. London: John Murray, 1892. P. 43.

277, Alhazen quote: Alhazen. Aporias against Ptolemy. As quoted in Sabra AI. Ibn al-Haytham: brief life of an Arab mathematician: died circa 1040. Harvard Magazine. 2003;September-October. Available at http://archive.ph/eYAIR.

277, Hume quote: Hume D. An enquiry concerning the principles of morals. Chicago: Open Court Publishing Co., 1912. P. 117. This book was first published in 1751.

277, second Ramon y Cajal quote: Ramon y Cajal S. Advice for a young investigator. Translated by Swanson N and Swanson LW. Cambridge, MA: MIT Press, 1999. P. 93. Translation based on the 4th edition of Ramon y Cajal's book published in 1916.

277, second Feynman quote: Feynman RP. What is science? Physics Teacher. 1969;7(September):320. From a lecture Feynman gave in 1966.

277, first Popper quote: Popper KR. The myth of a framework: in defence of science and rationality. Routledge: London, 1994. P. x. Popper wrote the section that includes this quote in 1993.

278, high confidence (a symptom of both deception and self-deception): See Trivers R. The folly of fools. New York: Basic Books, 2011.

282, Twain quote: Twain M. Mark Twain's notebook. New York: Harper and Brothers, 1935. P. 393.

282, Braben quote: Braben D. To be a scientist. Oxford: Oxford University Press, 1994. P. 143.

282, second **Popper** quote: Popper KR. The myth of the framework: in defence of science and rationality. Notturno MA (Editor). London: Routledge, 1994. P. 16. From a lecture Popper gave in 1973.

283, All scientists are prone to clinging to majorities: See Barber B. Resistance by scientists to scientific discovery. Science. 1961;134(3479):596–602.

284, extraordinary claims require extraordinary evidence: Carl Sagan coined this phrase (in episode 12, "Encyclopaedia Galactica," of the PBS television series Cosmos, which first aired on December 14, 1980), although David Hume (in *Philosophical Essays Concerning Human Understanding*, 1748) and Simon-Pierre Laplace (in *Théorie Analytique des Probabilités*, 1812) articulated similar ideas.

284, Lopez Corredoira quote: Lopez Corredoira M. The twilight of the scientific age. Boca Raton, Florida: Brown Walker Press, 2013. P. 142.

284, Clifford quote: Clifford WK. The ethics of belief. In The scientific basis of morals, and other essays. New York: Humboldt Publishing Company, 1884. P. 28. This essay was first published in 1877.

285, second **Payne-Gaposchkin** quote: Payne-Gaposchkin C. Cecilia Payne-Gaposchkin: an autobiography and other recollections, second edition. Edited by Haramundanis K. Cambridge: Cambridge University Press, 1996. P. 233. Payne-Gaposchkin first published this book in 1979.

286, Descartes quote: Descartes R. Discourse on method and the meditations. Translated by Veitch J. Amherst, NY: Prometheus

Books, 1989. P. 73. Descartes first published his *Meditations on First Philosophy* in 1641.

286, Feynman considered this extension of doubt to other realms one of the main values of science: Feynman RP. The value of science. In Feyman RP, What do you care what other people think? New York: Bantam Books, 1988. Pp. 240–248. From a lecture Feynman gave in 1955.

Chapter 62. Resisting Corruption

288, first **Solzhenitsyn** quote: Solzhenitsyn A. Nobel lecture in literature 1970. Available at https://perma.cc/ZL9Z-TE3R (screenshot view).

288, Sinclair quote: Sinclair U. I, candidate for governor: and how I got licked. Berkeley, California: University of California Press, 1994, p. 109. This book was first published in 1935.

288, Eisenhower quote: Eisenhower DD. Farewell address, January 17, 1961. Available at https://perma.cc/NMG4-FSNC.

288-289, Kealey quote: Kealey T. The economic laws of scientific research. Basingstoke: Macmillan Press Ltd., 1996. Pp. 332–333.

289, Anderson quote: Anderson PW. More and different: notes from a thoughtful curmudgeon. Singapore: World Scientific Publishing Co Pte Ltd; 2011. Pp. 163–165.

290, a free market in which customers can easily judge the quality of their offerings: Many industries operate in much less than free markets. Moreover, research-based improvements in products and services may not be apparent to customers if the improvements are slight or variable, take a long time to manifest, or require research to demonstrate. Medicine is one domain where these conditions usually apply.

290, departments are most likely to hire and promote scientists who attract the most grant funding: See chapter 19 in Smolin L. The trouble with physics. Boston: Houghton Mifflin Company, 2007.

293, second **Solzhenitsyn** quote: From a commencement address Solzhenitsyn gave in 1978. See https://perma.cc/F85Y-VZJH.

294-295, Feynman quote: Feynman RP. The pleasure of finding things out. Cambridge, MA: Perseus Publishing, 1999. P. 216. From a commencement address Feynman gave in 1974.

Appendix: Checklists

In these checklists, the numbers in square brackets refer to the chapter numbers of the corresponding chapter titles.

For the Pre-Study and Writing checklists, I suggest re-reading sets of chapters in the order I have listed. I recommend doing the corresponding practices in roughly the same order, if possible, although you can always do any of them again as necessary at any point.

Pre-Study (from conception through final preparation)

first set

Choosing Productive Research Questions [48]
No Substitute for Field Experience and Observation [2]
Searching the Literature [5]
Designing and Planning Research [3]
Hypothesis Evaluation, Description, and Exploration [51]
Overturning the Law of the Method [50]
Considering the Unseen [57]
Your are not Your Hypotheses or Results [53]
Doubting and Questioning [61]

second set

Keep it Simple, Scientist (KISS) [13]
Collecting Data [6]
Managing Data [7]
Analyzing Data [8]
Replicating Research [14]
Backing Up Research [12]

third set

Pre-Registering Research [4]
Soliciting Comments [31]

possibly relevant, depending on type of research or particulars of a project (all would be in first set):

Using Secondary Data [10] (secondary research/data analysis)
Reviewing Systematically [15] (systematic reviews, meta-analysis)
Doing Research without Funding [16] (unfunded/self-funded research)
Working with Other Researchers [35] (collaborative research)
Negotiating Authorship [39] (collaborative research)
Working with Research Partners [37] (research depending on persons/organizations beyond your research team)
No Substitute for Experimental Control [52] (research evaluating causal hypotheses)
Simulations are … Simulations [54] (modeling)

Writing a Research Report

first set

Simplifying the Writing Process [18]
Using Accurate Language [27]
Writing Across Generations [28]
Less is More [29]
Writing a Methods Section [19]
You are not Your Hypotheses or Results [53]
Figuring [25]
Writing a Results Section [20]
Citing Sources [26]
Reasoning [56]
Considering the Unseen [57]
Judging Hypotheses [58]
Doubting and Questioning [61]
Writing a Discussion Section [21]
Writing an Introduction [22]
Writing an Abstract [23]
Writing Acknowledgments [24]

Checking—Part II [30]

second set

Soliciting Comments [31]
Archiving and Sharing Data [9]
Publish, Then Perish [32]

Each of the following chapters correspond to **activities not tied to a particular research phase**. Consult these chapters in anticipation of or while doing these activities.

Programming [11] (if you do not yet program)
Giving a Scientific Presentation [33]
Dealing with the Press [34]
Contacting Other Scientists [40]
Responding to Other Scientists [41]
Engaging in Scientific Debates [43]
Recognizing Scientific Skills, Knowledge, and Character [36]
Managing vs. Doing Research [38]
Giving Comments [42]
Chairing a Conference Session [43]
Crossing Disciplinary Boundaries [49]

The following chapters focus on **routine or general purpose practices**. Consult these chapters occasionally until the practices become habitual, or when you need to refresh your memory.

Reasoning [56]
Considering the Unseen [57]
Chasing Wild Geese [45]
Making Notes [46]
Reading the Literature [47]
Judging Hypotheses [58]
Resisting Corruption [62]
Doubting and Questioning [61]

The following chapters cover **general principles** to keep in mind while doing and thinking about research.

Curbing Scientific Chauvinism [59]
Analysis for Understanding [55]
Getting Immediate Rewards from Doing Research [60]

Acknowledgments

I first had the idea for this book when Charles Replogle of Science Docs, Inc., invited me to write some blog posts on practical matters of doing scientific research. I soon realized that I had learned quite a few important lessons in my scientific career that researchers had rarely made explicit in print.

Many of my colleagues, clients, teachers, staff, students, and friends gave me the opportunities to learn these lessons and often counseled me wisely to avoid some of the pitfalls I describe. In particular, the following scientists have had particularly large influences on me in these ways: Ece Batchelder, the late Bill Batchelder, Russ Bernard, Stuart Brody, Simon Collery, Brad Dell, Mike Dell, Mark Fleisher, Charlie Fleming, the late Lin Freeman, Sharon Garrett, David Gisselquist, Matt Golden, Jeff Johnson, Jerry Kirk, Barbara Leigh, Fredrik Liljeros, Marc Miller, Steve Muth, John Potterat, John Roberts, Jr., Kim Romney, Rich Rothenberg, and Cindy Webster. I have also benefited greatly from many scientists, both living and dead, whom I have never met and do not know personally, but whose work and clear writing have served as beacons for me. Some of these hundreds of pioneers include David Buss, Dorothy Cheney, Robert Seyfarth, John Ioannidis, Thomas Kuhn, Robert Rosenthal, Thomas Sowell, and Peter Ward.

My mother, Carolyn Sabin, my late father, Jackson Brewer, and my brother, Britt Brewer, instilled in me the love for learning from an early age. I had many teachers from elementary school through college who made learning fun and interesting, especially in science, math, and English. Several teachers had big positive impacts on me, including Susan Bauer, Albert Black, Jr., Jody Brogan, Wendell Buck, Carol Eastman, Bob Eckenrode, Steve Harrell, George Kersul, Don Mackay, June Morita, Lois Neswick, Jack Raney, and Don "Buzzie" Welch. I am grateful to several of my athletics coaches, especially Gene Myers, who fostered discipline and honesty.

James Altucher inspired me to write and publish books myself. John Potterat showed me, by example, how fun and easy it is to write a book as a series of short chapters. My wife, Kathy Brewer, helped me identify some of the practices I recommend in this book, as they are not regularly used by non-scientists and can sometimes create conflict when used outside of scientific contexts. This book would not have been possible without her support, patience, and sacrifice. Discussions with my sons, Ned and Greg Brewer, also revealed to me essentials in research that are easy to overlook.

John Potterat and John Roberts, Jr., read and commented on drafts of every chapter. Stuart Brody, the late Linton Freeman, David Gisselquist, Al Klovdahl, Steve Muth, and Judy Van Raalte read and commented on drafts of a few chapters. Russ Bernard, Britt Brewer, Stuart Brody, David Gisselquist, Peter Lawrence, Thomas Nelson, and Irini Topalidou critiqued the penultimate version of the whole book. Together these reviewers found several errors and gaps, and gave valuable suggestions. I am very grateful for their help and feedback. John Potterat also kindly translated several quotes from French to English. Jay Fraser gave me helpful advice on book design and Britt Brewer gave me other useful feedback while I wrote the book.

Despite my debts to these named and other unnamed persons, I alone am responsible for the views in this book and any errors are mine. My acknowledgment of these individuals' influences does not mean they necessarily endorse my advice.

Request for Comments

Although I sought feedback on earlier versions of this book from many scientists before publication, I seek criticisms and suggestions for improvement from readers. I am particularly interested in learning about errors, omissions, and inconsistencies of any kind. Please contact me at www.evidenceguides.com.

Free Books for Libraries

Libraries are the only institution necessary for scientists to do research well. Therefore, I will give one print copy of this book free to any library open to the public that requests it. For details, please visit www.evidenceguides.com.

Getting Help

Some scientists lack skilled and experienced colleagues or others from whom they can get feedback or tutoring. I offer researchers of all levels personalized instruction and advice on scientific writing and the other topics I cover in this book. This includes online courses as well as individual tutoring. Go to www.writingforscientists.com for more information. Learning how to write well is much cheaper and more efficient than paying others to edit and write for you. It is also better than having little scientific impact due to poor writing and reasoning.

About the Author

Devon D. Brewer has 34 years of research experience in the health and social sciences. He has done research in academia, government, and the private sector, and also worked extensively as an independent scientist and consultant. Learn more about his research at www.interscientific.net.